Successful Writing at Work

CONCISE SECOND EDITION

CONCISE SECOND EDITION

Successful Writing at Work

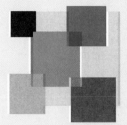

Philip C. Kolin
University of Southern Mississippi

HEINLE
CENGAGE Learning

Australia • Brazil • Japan • Korea • Mexico • Singapore • Spain • United Kingdom • United States

HEINLE
CENGAGE Learning

Successful Writing at Work,
Concise Second Edition
Philip C. Kolin

Executive Publisher: Patricia A. Coryell

Editor in Chief: Carrie Brandon

Senior Sponsoring Editor: Lisa Kimball

Senior Marketing Manager: Tom Ziolkowski

Discipline Product Manager: Giuseppina Daniel

Senior Project Editor: Margaret Park Bridges

Senior Media Producer: Philip Lanza

Senior Content Manager: Janet Edmonds

Art and Design Manager: Jill Haber

Cover Design Manager: Anne S. Katzeff

Senior Photo Editor: Jennifer Meyer Dare

Senior Composition Buyer: Chuck Dutton

New Title Project Manager: Susan Peltier

Editorial Associate: Sarah Truax

Marketing Assistant: Bettina Chiu

Cover image: © Cover illustration by
   Mick Wiggins

To Kristin, Eric, and Theresa
Evan Philip and Megan
Elise
Julie and Loretta
and
MARY

For product information and technology assistance, contact us at
Cengage Learning Customer & Sales Support, 1-800-354-9706

For permission to use material form this text or product,
submit all requests online at **www.cengage.com/permissions**
Further permissions questions can be emailed to
permissionrequest@cengage.com

Library of Congress Control Number: 2007931299
   ISBN-10: 0-618-94864-3
   ISBN-13: 978-0-618-94864-2

Heinle
25 Thomson Place
Boston, MA 02210
USA

Cengage Learning is a leading provider of customized learning solutions with office location around the globe, including Singapore, the United Kingdom, Australia, Mexico, Brazil and Japan. Locate your local office at: **international.cengage.com/region**

Cengage Learning products are represented in Canada by Nelson Education, Ltd.

For your course and learning solutions, visit **academic.cengage.com**
Visit Heinle online at **elt.heinle.com**
Visit our corporate website at **www.cengage.com**

Printed in the United States of America
6 7 11 10

CONTENTS

PART II: Correspondence

Chapter 3: Writing Memos, Faxes, and E-Mails

Chapter 4: Writing Letters

Chapter 5: How to Get a Job: Résumés, Letters of Application, and Interviews

PART III: Preparing Documents and Visuals

Chapter 6: Designing Successful Documents and Visuals   168

PREFACE

Overview

Successful Writing at Work, Concise Second Edition, is a practical introductory text for students in business, professional, and occupational writing courses. As readers of the full-length edition of this text have found, *Successful Writing at Work* clearly helps students develop and master key communication skills vital for success in the workplace. The *Concise Edition* serves the same purpose, but it is designed for those who prefer a more compact text, one that covers nearly as many business writing topics but is more streamlined and focuses on the most essential skills and strategies for writing successfully on the job. Whereas the full-length edition includes seventeen chapters, the *Concise Edition* contains ten chapters, yet fully covers the range of workplace communication: from essential considerations such as audience analysis and ethics, to writing increasingly more complex business documents (memos through long reports), to making presentations, to preparing a résumé and interviewing for a job.

This compact edition has been designed for a variety of educational settings where business writing is taught. It is versatile enough for a full semester or trimester course, or it can be used successfully in a shorter course, such as on a quarter system. It can also meet the diverse goals of varied educational settings, including online, distance education, continuing education, and week-long intensive courses, as well as in-house training programs, workshops, and conferences.

Successful Writing at Work, Concise Second Edition, provides students with easy-to-understand guidelines for writing and designing clear, well organized, and readable documents. Along with user-friendly guidelines, this edition provides students with realistic models of the precise kinds of documents they will be asked to write on the job. In addition, this text can serve as a ready reference that readers can easily carry with them to the workplace. Students will quickly find that this book includes many practical applications, which are useful to those who have little or no job experience as well as those with years of experience in the world of work.

Approaches

The *Concise Second Edition* integrates the following approaches to teaching occupational writing throughout the book:

1. **Focus on writing as a problem-solving activity.** The book approaches business writing not merely as a set of rules and formats, but simultaneously as a problem-solving activity that helps employees meet the needs of their employers,

co-workers, customers, clients, community groups, and vendors. This approach to writing, as introduced in Chapter 1 and carried throughout the text, asks students to think through the writing process, asking the questions *who* (who is the audience?), *why* (why do they need this document?), *what* (what is the message?), and *how* (what is the style and tone?), rather than offering inflexible prescriptions for what constitutes effective business communication.

2. **Emphasis on audience analysis, including international audiences and non-native speakers of English.** A key feature of the *Concise Second Edition* is its consistent emphasis on the first problem-solving question: *Who is the audience?* This question, discussed in depth in Chapter 1, affects everything else that goes into the writing of a workplace document, from determining its purpose to creating its message, to selecting its style and tone. All subsequent chapters stress how the needs and expectations of the audience must be considered alongside the scope and format of various types of documents. In response to the needs of today's workplace, the text expands the discussion of audience to include international readers, beginning in Chapter 1 and carried out through the book.

3. **View of student readers as business professionals.** To encourage students in their job-related writing, the *Concise Second Edition* treats them as professionals seeking success at different phases of their business careers. Students are asked to place themselves in the workplace setting (or in the case of Chapter 5, in the role of job seekers) as they approach each topic, to understand the differences between workplace and academic writing. Students are asked to see themselves as collaborative employees (Chapter 2), workers writing routine documents (Chapters 3 and 4), employees designing and writing more complex documents and reports (Chapters 6–9), and as company representatives designing and making presentations (Chapter 10).

4. **Inclusion of the most current workplace technologies.** The text shows students how to become better problem solvers and writers by using the varied resources of an evolving, ever more complex technological workplace. In light of these expanding communication resources, the *Concise Second Edition* helpfully discusses technological considerations alongside the guidelines to writing print documents. In every chapter, easy-to-understand explanations assist students in discovering the *hows* as well as the *whys* of writing for the digital world of work.

New to the Concise Second Edition

As in the first *Concise Edition*, the second edition continues to offer students a streamlined alternative to the full edition of *Successful Writing at Work* while still providing many important additions. Highlights of the new edition include:

- **Expanded coverage of writing for international readers and non-native speakers of English.** Several new sections have been added on these crucial topics. A new section entitled "Writing for the Global Marketplace" in Chapter 1, "Getting Started: Writing and Your Career," emphasizes the importance of international audiences and non-native speakers of English in the American work force. In Chapter 4, "Writing Letters," coverage of writing letters for international readers has been expanded. Chapter 6, "Designing Successful Documents and Visuals," features a new section on using appropriate visuals for international audiences. Chapter 9, "Writing Careful Long Reports," focuses on the role international workers play in a global corporation. Chapter 10, "Making Successful Presentations at Work," includes a new section on making presentations for international audiences. Finally, the website now includes several Additional Activities on the topic of writing for international readers.

- **New section on collaborative writing and expanded collaborative exercises.** The new collaborative writing section in Chapter 2, "The Writing Process and Collaboration at Work," includes an overview of collaboration in the workplace setting, guidelines for collaborative writing, ways to troubleshoot group conflicts, and a section on online technologies for collaboration with a case study to reinforce the steps. In addition, exercises throughout the book have been updated to include new collaborative assignments, and the website now includes new collaborative Additional Activities.

- **Enhanced coverage of ethics.** This new edition pays even closer attention to ethics in workplace writing than the previous edition. In addition to the book's emphasis on the ethical considerations of writing for international readers and non-native speakers of English, Chapter 1, "Getting Started: Writing and Your Career," features a new section on ethical dilemmas, and Chapter 6, "Designing Successful Documents and Visuals" contains an enhanced discussion of constructing, inserting, and writing about visuals ethically.

- **Updated coverage of workplace technologies.** To keep up with the ever-changing digital workplace, technology coverage has been updated and expanded. Chapter 2, "The Writing Process and Collaboration at Work," now includes new material on collaborating online; Chapter 3, "Writing Memos, Faxes, and E-Mails," features a new section on e-mails as legal records; Chapter 5, "How to Get a Job: Résumés, Letters of Application, and Interviews," provides an expanded section on online résumés with an annotated example; and technology is woven into the writing instruction throughout the text.

- **Updated and expanded visuals chapter.** Chapter 6 has been retitled "Designing Successful Documents and Visuals" to de-emphasize Web design (now removed) and expand upon such topics as using color, inserting clip art, the ethics of visuals, and the appropriate use of visuals for international readers.

- **Revised report in the long reports chapter.** The long report in Chapter 9, "Adapting the U.S. Workplace for Multinational Employees," has been substantially revised to better illustrate the type of report employees will have to prepare in the business world. The report reflects the content, format, organization, and documentation business readers will expect to see in a long report.

- **New PowerPoint presentation in the chapter on oral communication at work.** Chapter 10, "Making Successful Presentations at Work," now includes an annotated PowerPoint presentation.

- **Complete annotations of model documents.** Every document in the text has been thoroughly annotated to better illustrate specifically how ineffective documents can be improved and how effective documents and visuals are constructed with the principles of good workplace writing in mind.
- **Updated figures and exercises.** The figures have been updated throughout the new edition not only for currency but also to show students the importance of effective, up-to-date graphics in their own work. In addition, new exercises on the use of visuals have been added throughout this new edition.

Supplemental Resources for Students and Instructors

The Online Study Center is Cengage Learning's comprehensive location for extensive interactive online products and services to accompany composition texts. Students and instructors can access the Online Study Center content through text-specific student and instructor websites; via Eduspace®, Cengage Learning's online course management system; and through other course management systems, including Blackboard and WebCT. For a demonstration of the Online Study Center or to discuss special packaging options available for this text, please consult your local Cengage Learning representative (locatable through *academic.cengage.com*).

- **The Online Study Center Instructor Website (college.cengage.com/pic/ kolinconcise2e)**

Instructors can access the Online Study Center content at any time via the Internet. Resources include chapter-by-chapter PowerPoint slides, sample syllabi, and teaching suggestions.

- **The Online Study Center Student Website (college.cengage.com/pic/ kolinconcise2e)**

Accessible to students via the Internet, The Online Study Center content for this text includes additional activities, web links, ACE self-tests, and guidance on documentation, including a comparison of MLA and APA documentation styles. Some content may be passkey-protected.

- **The Online Study Center with Eduspace**

Eduspace, Cengage Learning's course management system, offers instructors a flexible and interactive online platform for communicating with students, organizing course material, evaluating student work, and tracking results in a powerful gradebook. In addition to the Online Study Center resources, students and instructors using Eduspace benefit from course management tools, including a powerful gradebook and practice and homework exercises.

▌Acknowledgments

In a very real sense, the *Concise Second Edition* has profited from the collaboration of various reviewers with the author. I am, therefore, honored to thank the following reviewers who have joined with me to create this new edition.

Sonya Compton Borton, *University of Louisville*

Kristin Dietsche, *Northern Kentucky University*

Scott Downing, *DePaul University*

Ronald G. Mullins, *Bronx Community College*

Cynthia Murrell, *New Mexico State University*

Andrea Penner, *San Juan College*

Lourdes Rassi, *Florida International University*

Leticia Slabaugh, *Arizona State Univesity – Tempe*

David R. Swarts, *Clinton Community College*

I am also deeply grateful to the following individuals at the University of Southern Mississippi for their help: Beverly Ciko, Sherry Smith, and my chair, Michael Mays (Department of English), Mary Lux (Department of Medical Technology), and Kay Wall (Cook Library).

Several individuals from business and industry also gave me valuable assistance, for which I am thankful. They include Joycelyn Woolfolk at the Federal Reserve Bank in Atlanta, Sally Eddy at Georgia Pacific, John Krumpos at Gulf Paper Company, Hilary J. Englert at Rice's Potato Chips, Don McCarthy (an independent computer programmer), Russel Dukette at Petro Automotive Group, and Dr. Matt Fry and Dr. Michael O'Neal from the Hattiesburg Clinic.

I appreciate the input of the following two individuals, who provided the Chinese and Turkish translations that appear in Chapter 7: Jian Zheng and Fatih Uzuner.

I am also especially grateful to Father Michael Tracey for his counsel and his contributions to Chapter 6 on document design.

My thanks go to my editors at Heinle, Cengage Learning for their assistance, encouragement, and friendship: Lisa Kimball, Bruce Cantley, Charline Lake, Margaret Bridges, Sarah Truax, and Katilyn Crowley.

Finally, I am grateful to my son Eric, my daughter-in-law Theresa, my grandson Evan Philip, and my granddaughter Megan Elise for their love and encouragement. My daughter Kristin Julie merits extra praise for the many times she assisted me by doing various searches and revisions.

P.C.K.

January 2008

Successful Writing at Work

CONCISE SECOND EDITION

Getting Started

Writing and Your Career

Writing—An Essential Job Skill

Writing is a part of every job. In fact, your first contact with a potential employer is through posting your résumé and writing a letter of application, which determine a company's first impression of you. And the higher you advance in an organization, the more writing you will do. Promotions are often based on a person's writing skills.

The Associated Press reported in a recent survey that "most American businesses say workers need to improve their writing . . . skills." The same report cited a survey of 402 companies that identified writing as "the most valued skill of employees." Still, the employers polled in that survey indicated that 80 percent of their employees need to improve their writing skills. According to Don Bagin, a communications consultant, most people need an hour or more to write a typical business letter. If an employer is paying someone $30,000 a year, one letter costs $14 of that employee's time; for someone who earns $50,000 a year, the cost of the average letter jumps to $24. Clearly, writing is an essential skill for anyone in the workplace—employers and employees alike.

This chapter gives you some basic information about writing in the global marketplace and offers major questions you can ask yourself to make the writing process easier and the results more effective. It also describes the basic functions of on-the-job writing and introduces you to one of the most important requirements in the business world—writing ethically.

Writing for the Global Marketplace

The Internet, e-mail, weblogs, teleconferencing, and e-commerce have shrunk the world into a global village. It is no longer feasible to think of business in exclusively regional or even national terms. Many companies are multinational corporations with offices throughout the world. A large, multinational corporation may have its equipment designed in Japan, built in Bangladesh, and sold in Detroit, Atlanta, and

To expand your understanding of writing and your career, take advantage of the Web Links, Additional Activities, and ACE Self-Tests at **college.cengage .com/pic/ kolinconcise2e.**

Chapter 1 Additional Activities, located at **college.cengage .com/pic/ kolinconcise2e,** include a global-focused activity titled "Exploring the Global Marketplace."

1

Los Angeles. In this global economy, every country is affected by every other one, and all of them are connected by the Internet.

Every business, whether large or small, has to appeal to diverse international markets to be competitive. Each year a larger share of the U.S. gross national product (GNP) depends on global markets. Some U.S. firms estimate that 30 to 40 percent of their business is conducted outside the United States. Wal-Mart, for example, operates hundreds of stores in mainland China, and General Electric has plants in fifty countries. If your company has a website, then it is an international business.

To be a successful employee in this highly competitive, global market, you will have to communicate clearly and diplomatically with a host of readers from different cultural backgrounds. As a result, don't presume that you will be writing only to native speakers of American English. As a part of your job, you may communicate with readers in Singapore, Malaysia, or India, for example, who speak varieties of English quite different from American English. You will also very likely be writing to readers for whom English is not their first (or native) language. These individuals, who may reside either in the United States or in a foreign country, constitute a large and important audience for your work.

See the World Through Your Reader's Eyes

Writing to international readers with proper business etiquette means first learning about their cultural values and assumptions—what they esteem and also what they regard as communication taboos. They may not do business exactly the way it is done in the United States, and to think they should is wrong. Your international audience is likely to have different expectations of how they want a letter addressed or written to them, how they prefer a proposal to be submitted, or how they wish a business meeting to be conducted. Their concepts of time, family, money, work, managers, and communication itself may be nothing like what they are in the United States. If you misunderstand their culture by inadvertently writing or saying something inappropriate, it can cost your company a contract and you your job.

Cultural diversity exists inside as well as outside the company you work for. Don't conclude that your boss or co-workers are all native speakers of English, either, or that they come from the same cultural background that you do. In the next decade, as much as 40 to 50 percent of the U.S. skilled work force may be composed of immigrants who bring their own traditions and languages with them. For the common good of your company, you need to be respectful of these international colleagues. The long report on pages 117–125 describes some ways in which you can both acknowledge and respect the different cultural traditions of these colleagues in the world of work. Businesses want to emphasize their international presence. Figure 1.1, for example, shows how a large corporation like Citibank helps customers worldwide.

Use International English

Whether these international readers are your customers or co-workers, you will have to adapt your writing to respect their language needs and communication pro-

A company's dedication to globalization.   **Figure 1.1**

How Citigroup Meets Banking Needs Around the World

WITH A BANKING EMPIRE that spans more than 100 countries, Citigroup is experienced at meeting the diverse financial services needs of businesses, individuals, customers, and governments. The bank is headquartered in New York City but has offices in Africa, Asia, Central and South America, Europe, the Middle East, as well as throughout North America. Live or work in Japan? You can open a checking account at Citigroup's Citibank branch in downtown Tokyo. How about Mexico? Visit a Grupo Financiero Banamex-Accival branch, owned by Citigroup. Citigroup owns European American Bank and has even bought a stake in a Shanghai-based bank with an eye toward attracting more of China's $1 trillion in bank deposits. Between acquisitions and long-established branches, Citigroup covers the globe from the Atlantic to the Pacific and the Indian Oceans.

Individuals can use Citigroup for all the usual banking services . . . . Personalized service is the hallmark of . . . the bank, which can help prepare customized financial plans for . . . customers, manage their securities trading activities, provide trust services, and much more. What's more, Citigroup is active in communities around the world through philanthropic contributions and grants, financial literacy seminars, volunteerism, and supplier diversity programs. This financial services giant strives for the best of both worlds, wielding its global presence and resources to meet banking needs locally, one customer at a time.

tocols. The words, idioms, phrases, and sentences you choose instinctively for U.S. readers may not be appropriate for an audience for whom English is a second, or even third, language. If you find a set of directions confusing, imagine how much more intimidating such a document would be for a non-native speaker of English. To communicate with non-native speakers, you will have to use "international English," a way of writing that is easily understood, culturally tactful, and diplomatic. International English simply means that your message is clear, straightforward, and appropriate for readers who are not native speakers of English. It is free from complex, hard-to-process sentences as well as from cultural biases. International English is discussed in detail on pages 117–120. Later chapters of this book also give you guidelines on writing e-mail, correspondence, instructions, reports, websites, and other work-related documents for the global marketplace.

Four Keys to Effective Writing

Effective writing on the job is carefully planned, thoroughly researched, and clearly presented. Its purpose is always to accomplish a specific goal and be as persuasive as possible. Whether you send a routine e-mail to a co-worker in Cincinnati or in China or a special report to the president of the company, your writing will be more effective if you ask yourself four questions.

1. *Who* will read what I write? (Identify your *audience*.)
2. *Why* should they read what I write? (Establish your *purpose*.)
3. *What* do I have to say to them? (Formulate your *message*.)
4. *How* can I best communicate? (Select your *style* and *tone*.)

The questions *who? why? what?* and *how?* do not function independently; they are all related. You write (1) for a specific audience (2) with a clearly defined purpose in mind (3) about a topic your readers need to understand (4) in language appropriate for the occasion. Once you answer the first question, you are off to a good start toward answering the other three. Now let us examine each of the four questions in detail.

Identifying Your Audience

Knowing *who* makes up your audience is one of your most important responsibilities as a writer. Expect to analyze your audience throughout the composing process.

Look for a minute at the posters in Figures 1.2, 1.3, and 1.4. The main purpose of all three posters is the same: to discourage people from smoking. The essential message in each poster—smoking is dangerous to your health—is also the same. But note how the different details—words, photographs, situations—have been selected to appeal to three different audiences.

The poster in Figure 1.2 is aimed at fathers who smoke. As you can see, it is an image of a father smoking next to his son, who is reaching for his pack of cigarettes. Note how the caption "Will Your Child Follow in Your Footsteps?" plays on the fact that the father and son are literally sitting on steps, but at the same time implies that the son will imitate the father's behavior as a smoker. The statistic at the bottom

No-smoking poster aimed at fathers who smoke.   **Figure 1.2**

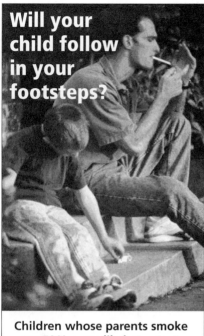

Photo by Peter Poulides/Getty Images

of the poster reinforces both the photo caption and the image, hitting home the point that parental behavior strongly influences children's behavior. Already the child in the photograph is following his father by showing a clear interest in smoking.

The poster in Figure 1.3, on the other hand, is aimed at an audience of pregnant women and shows a pregnant woman with a lit cigarette. The words on the poster appeal to a mother's sense of responsibility as the reason to stop smoking, a reason to which pregnant women would be most likely to respond.

Figure 1.4 is directed toward young athletes. The word *smoke* is aimed directly at their game and their goal. In fact, the poster writer appropriately made the goal the same for the game as for the players' lives. Note, too, how this image is suitable for an international audience.

The copywriters who created these posters have chosen appropriate details—questions, pictures, captions, and so on—to convince each audience not to smoke. With their careful choices, they successfully answered the question, "How can we best communicate with each audience?" Details relevant for one audience (athletes, for example) could not be used as effectively for another audience (such as fathers).

Figure 1.3   No-smoking poster directed at pregnant women.

Photo by Bill Crump/Brand X Pictures/Fotosearch/Royalty-Free Image

Keep the following in mind:

- Members of each audience differ in backgrounds, experiences, needs, and opinions.
- How you picture your audience will determine what you say to them.
- Viewing something from the audience's perspective will help you to select the most relevant details for that audience.

Some Questions to Ask About Your Audience

You can form a fairly accurate picture of your audience by asking yourself some questions *before* you write. For each audience for whom you write, consider the following questions.

1. **Who is my audience?** What individual(s) will most likely be reading my work?

   If you are writing for individuals at work:

   - What is my reader's job title? Co-worker? Immediate supervisor? Vice president?
   - What kind of job experience, education, and interests does my reader have?

No-smoking poster appealing to young athletes.   **Figure 1.4**

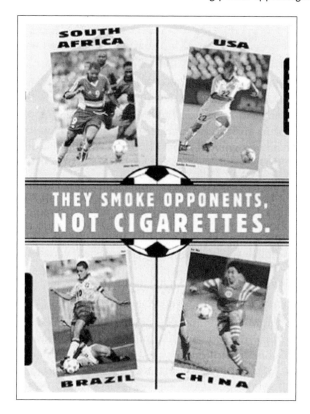

If you are writing for clients or consumers (a very large, sometimes fragmented audience):

- How can I find out about their interest in my product or service?
- How much will this audience know about my company? About me?

2. **How many people will make up my audience?**

- Will just one individual read what I write (the nurse on the next shift, the production manager) or will many people read it (all the consumers of a product)?
- Will my boss want to see my work (say, a letter to a consumer in response to a complaint) to approve it?
- Will I be sending my message to a large group of people sharing a similar interest in my topic, such as a listserv?

3. **How well does my audience understand English?**

- Are all my readers native speakers of English?
- Will I be communicating with people all around the globe?

- Will some of my readers use English as a second or even third language, and so require extra sensitivity on my part to their needs as non-native speakers of English?
- Will some of my readers speak no English, but instead use an English grammar book and foreign language dictionary to understand what I've written?

4. How much does my audience already know about my topic?

- Will my audience know as much as I do about the particular problem or issue, or will they need to be briefed or updated?
- Are my readers familiar with, and do they expect me to use, technical terms and descriptions, or will I have to provide easy-to-understand comparisons and nontechnical summaries?

5. What is my audience's reason for reading my work?

- Is reading my communication part of their routine duties, or are they looking for information to solve a problem or make a decision?
- Am I writing to describe benefits that another writer or company cannot offer?
- Will my readers expect complete details, or just a short summary?
- Are they reading my work to make an important decision affecting a co-worker, a client, or a community?
- Are they reading something I write because they must (a legal notification, for instance)?

6. What are my audience's expectations about my written work?

- Do they want a brief e-mail or will they expect a formal letter?
- Will they expect me to follow a company format and style?
- Are they looking for a one-page memo or for a comprehensive report?
- Should I use a formal tone or a more relaxed and conversational style?

7. What is my audience's attitude toward me and my work?

- Will I be writing to a group of disgruntled customers and/or vendors about a sensitive issue (a product recall, a refusal of credit, or a shipment delay)?
- Will I have to be sympathetic while at the same time give firm reasons for my company's (or my) decision?
- Will my readers be skeptical, indifferent, or friendly about what I write?
- Will my readers feel guilty that they have not answered an earlier message of mine, not paid a bill now overdue, or not kept a promise or commitment?

8. What do I want my audience to do after reading my work?

- Do I expect my reader to purchase something from me, approve my plan, or send me additional materials?
- Do I simply want my reader to get my message and not respond at all?
- Do I expect my reader to get my message, acknowledge it, save it for future reference, or review it and e-mail it to another individual or office?
- Does my reader have to take immediate action, or does he or she have several days or weeks to respond?

As your answers to these questions will show, you may have to communicate with many different audiences on your job. If you work for a large organization with numerous departments, you may have to write to such diverse readers as accountants, iTech staff, engineers, public relations specialists, marketing experts, programmers, and individuals who install, operate, and maintain equipment. On some occasions, you may have to write for multiple audiences at once—a manager, a buyer, and someone in your firm's legal department. In addition, you need to communicate effectively with customers about your company's products and services.

Each group of readers has different expectations and requirements; you need to understand those audience differences if you want to supply relevant information. Let's say you work for a manufacturer of heavy-duty equipment and that you have to write for many individuals in an organization that relies on such equipment. Here are the priorities of five different audiences and selected appropriate information to communicate with each one.

Audience	Information to Communicate
Owner or principal executive	The writer would stress financial benefits, indicating that the machine is a "money-maker" and is compatible with other existing equipment.
Production engineer	The writer needs to emphasize "state-of-the-art" transmissions, productivity, upkeep.
Operator	The writer would focus on information about how easy and safe it is to run the equipment.
Maintenance worker	The writer must provide key details about routine maintenance as well as troubleshooting advice on problems.
Production supervisor	The writer is expected to address the speed and efficiency the machine offers.

To succeed in the world of work, give each reader the details he or she needs to accomplish a given job.

Establishing Your Purpose

By knowing *why* you are writing, you will communicate better and find writing itself to be an easier process. Make sure you follow the most important rule in occupational writing: **Get to the point right away.** At the start of your message, state your goal clearly.

> I want to teach new employees the security code for logging on to the company computer.

Since your purpose controls the amount and order of information you include, state it clearly at the beginning of every e-mail, memo, letter, and report.

> This e-mail will acquaint new employees with the security measures they must take when logging on to the company computer.

In the opening purpose statement that follows, note how the author clearly informs the reader as to what the report will and will not cover.

> As you requested at last week's organizational meeting, I have conducted a study of our use of the Internet to advertise our services. This report describes, but does not evaluate, our current practices.

Formulating Your Message

Your message is the sum of *what* facts, responses, and recommendations you put into writing. A message includes the scope and details of your communication.

- *Scope* refers to how much information you give readers about the key details.
- The *details* are those key points you think readers need to know to perform their jobs.

Some messages will consist of one or two sentences: "Do not touch; wet paint." "Order #756 was sent this afternoon by Federal Express. It should arrive at your office on March 22." At the other extreme, messages may extend over twenty or thirty pages. Messages can carry good news or bad news. They may deal with routine matters; or they may handle changes in policy, special situations, or problems.

Keep in mind that you will adapt the message to fit your audience. For technical audiences, such as engineers or technicians, you may have to supply a complete report with every detail noted or contained in an appendix. For other readers—busy executives, for example—a short discussion or summary of the financial or managerial significance will be enough.

Selecting Your Style and Tone

Style

Style is *how* something is written rather than what is written. Style helps to determine how well you communicate with an audience, how well your readers understand and receive your message. It involves the choices you make about

- the construction of your paragraphs
- the length and patterns of your sentences
- your choice of words

You will have to adapt your style to take into account different messages, different purposes, and different audiences. The words, for example, will certainly vary with your audience. If all potential readers are specialists in your field, you may safely use the technical language and symbols of your profession. Nonspecialists, however, will be confused and annoyed if you write to them in the same way. The average consumer, for example, will not know what a *potentiometer* is; by writing "volume control on a radio," you will be using words that the general public can understand.

Tone

Tone in writing, like tone of voice, expresses your attitude toward a topic and toward your audience. In general, your tone can range from formal and impersonal (a scientific report) to informal and personal (e-mail to a friend or a how-to article for consumers). It can be unprofessionally sarcastic or diplomatically agreeable.

Tone, like style, is signaled in part by the words you choose. For example, saying that someone is "interested in details" conveys a more positive tone than saying the person is a "nitpicker." The word *economical* is more positive than *stingy* or *cheap; assertive* sounds better than *rude*.

The tone of your writing is especially important in occupational writing because it reflects the image you project to readers and thus determines how they will respond to you, your work, and your company. Depending on your tone, you can appear sincere and intelligent or angry and uninformed. Of course, in all written work, you need to sound professional and knowledgeable about the topic and genuinely interested in your readers' opinions and problems.

Style and Tone Examples

A Description of Heparin for Two Different Audiences

To better understand the effects of style and tone on writing, read the two excerpts on page 36. In both, the message is basically the same, but because the audiences differ, so do the style and the tone. The two pieces are descriptions of *heparin*, a drug used to prevent blood clots.

Technical/Scientific Style and Tone   The first description appears in a reference work for physicians and other health care providers and is written in a highly technical style with an impersonal tone.

The writer made the appropriate stylistic choices for the audience, the purpose, and the message. Health care providers understand and need the technical vocabulary the writer uses; this audience also requires the sophisticated and lengthy explanations to prescribe and/or administer heparin correctly. The author's authoritative, impersonal tone is coldly clinical, which, of course, is also correct because the purpose is to convey the accurate, complete scientific facts about this drug, not the writer's or reader's opinions or beliefs. The author sounds both knowledgeable and appropriately objective.

HEPARIN SODIUM INJECTION, USP
STERILE SOLUTION
Description: Heparin Sodium Injection, USP, is a sterile solution of heparin sodium derived from bovine lung tissue, standardized for anticoagulant activity.

Each ml of the 1,000 and 5,000 USP units per ml preparations contains: heparin sodium 1,000 or 5,000 USP units; 9 mg sodium chloride; 9.45 mg benzyl alcohol added as preservative. Each ml of the 10,000 USP units per ml preparations contains: heparin sodium 10,000 units; 9.45 mg benzyl alcohol added as preservative.

When necessary, the pH of Heparin Sodium Injection, USP, was adjusted with hydrochloric acid and/or sodium hydroxide. The pH range is 5.0 to 7.5. *Clinical pharmacology:* Heparin inhibits reactions that lead to the clotting of blood and the formation of fibrin clots both *in vitro* and *in vivo*. Heparin acts at multiple sites in the normal coagulation system. Small amounts of heparin in combination with antithrombin III (heparin cofactor) can inhibit thrombosis by inactivating activated Factor X and inhibiting the conversion of prothrombin to thrombin.

> *Dosage and administration:* Heparin sodium is not effective by oral administration and should be given by intermittent intravenous injection, intravenous infusion, or deep subcutaneous (intrafrat, i.e., above the iliac crest or abdominal fat layer) injection. **The intramuscular route of administration should be avoided because of the frequent occurrence of hematoma at the injection site.**[1]

Nontechnical Style and Tone   The following description of heparin, on the other hand, is written in a nontechnical style and with an informal, caring tone. This description is similar to those found on information cards given to patients about the drugs they are receiving in a hospital.

> Your doctor has prescribed a drug called *heparin* for you. This drug will prevent any new blood clots from forming in your body. Since heparin cannot be absorbed from your stomach or intestines, you will not receive it in a capsule or tablet. Instead, it will be given into a vein or the fatty tissue of your abdomen. After several days, when the danger of clotting has passed, your dosage of heparin will be gradually reduced. Then another medication you can take by mouth will be started.

The writer of the second description also made the appropriate choices for the readers and their needs. Familiar words rather than technical ones are suitable for nonspecialists such as patients. This audience does not need elaborate descriptions of the origin and composition of the drug. The tone is both personal and straightforward because the purpose is to win the patient's confidence and to explain the essential functions of the drug.

Characteristics of Job-Related Writing

Job-related writing characteristically serves six basic functions: (1) to provide practical information, (2) to give facts rather than impressions, (3) to provide visuals to clarify and condense information, (4) to give accurate measurements, (5) to state responsibilities precisely, and (6) to persuade and offer recommendations. These functions tell you what kind of writing you will produce after you successfully answer the *who? why? what?* and *how?*

Providing Practical Information

On-the-job writing requires a practical here's-what-you-need-to-do-or-to-know approach. One such practical approach is *action-oriented.* You instruct the reader to do

---

[1]Copyright © *Physicians' Desk Reference*® 45th edition, 1991, published by Medical Economics, Montvale, New Jersey 07645. Reprinted by permission. All rights reserved.

something—assemble a ceiling fan, test for bacteria, perform an audit, create a website. Another practical approach is *knowledge-oriented*: to have someone understand something—why a procedure was changed, what caused a problem or solved it, how much progress occurred on a job site, why a new piece of equipment should be purchased. Examples of knowledge-oriented practical writing are a letter from a manufacturer to customers to explain a product recall and an e-mail to employees about changes in their group health insurance.

The following description of Energy Efficiency Ratio combines both the action-oriented and knowledge-oriented approaches of practical writing.

> Whether you are buying window air-conditioning units or a central air-conditioning system, consider the performance factors and efficiency of the various units on the market. Before you buy, determine the Energy Efficiency Ratio (EER) of the units under consideration. The EER is found by dividing the BTUs (units of heat) that the unit removes from the area to be cooled by the watts (amount of electricity) the unit consumes. The result is usually a number between 5 and 12. The higher the number, the more efficiently the unit will use electricity.[2]

Giving Facts, Not Impressions

Occupational writing is concerned largely with those things that can be seen, heard, felt, tasted, or smelled. The writer uses *concrete language* and specific details. The emphasis is on facts rather than on the writer's feelings or guesses.

The discussion on the next page by a group of scientists about the sources of oil spills and their impact on the environment is an example of writing with objectivity. It describes events and causes without anger or tears.

> Major oil spills occur as a result of accidents such as blowout, pipeline breakage, etc. Technological advances coupled with stringent regulations have helped to reduce the chances of such major spills; however, there is a chronic low-level discharge of oil associated with normal drilling and production operations. Waste oils discharged through the river systems and practices associated with tanker transports dump more significant quantities of oils into the ocean, compared to what is introduced by the offshore oil industry. All of this contributes to the chronic low-level discharge of oil into world oceans. The long-range cumulative effect of these discharges is possibly the most significant threat to the ecosystem.[3]

Providing Visuals to Clarify and Condense Information

Visuals are indispensable partners of words in conveying information to your readers. On-the-job writing makes frequent use of tables, charts, photographs, flow charts, diagrams, and drawings. Chapter 6 discusses the use of visuals.

Visuals play an important role in the workplace. Note how the drawing in Figure 1.5 can help computer users better understand and follow the accompanying written guidelines. A visual like this, reproduced in an employee handbook or displayed on a website, can significantly reduce physical stress and increase productivity.

---

[2]Reprinted by permission of Entergy New Orleans.
[3]*The Offshore Ecology Investigation*, Galveston: Gulf Universities Research Consortium.

Figure 1.5   Use of a visual to convey information.

An Ergonomic Workstation

Ergonomics is the science of designing workplace equipment to conform to human needs. To maintain comfort while you work and avoid workplace injuries, follow the principles shown in the photograph and discussed in the accompanying text below.

- To reduce the possibility of eye damage, maintain a distance of 18 to 24 inches between your eyes and the computer screen and keep your work area well lit.
- To reduce neck strain, position your computer screen so that the top of the screen is at or just below eye level.
- To avoid back and shoulder stress, sit up straight in your chair with your shoulders relaxed and your lower back firmly supported (with a cushion, if necessary).

- To prevent carpal tunnel syndrome and arm strain, rest your elbows at your side and keep your forearms and wrists level with the keyboard.
- To avoid bumping your thighs and knees, allow 1–2 inches of space between your thighs and the keyboard shelf.
- To prevent leg and back strain, adjust your chair height so that your upper and lower legs form a 90-degree angle and your feet are either flat on the floor or on a footrest.

Photo courtesy Ergo Concepts, LLC

Giving Accurate Measurements

Much of your work will depend on measurements—acres, bytes, calories, centimeters, degrees, dollars and cents, grams, percentages, pounds, square feet, units. Numbers are clear and convincing. The following discussion of mixing colored cement for a basement floor would be useless to readers if it did not supply accurate quantities.

> The inclusion of permanent color in a basement floor is a good selling point. One way of doing this is by incorporating commercially pure mineral pigments in a topping mixture placed to a 1-inch depth over a normal base slab. The topping mix should range in volume between 1 part Portland cement, 1¼ parts sand, and 1¼ parts gravel or crushed stone and 1 part Portland cement, 2 parts sand, and 2 parts gravel or crushed stone. Maximum size gravel or crushed stone should be ⅜ inch.[4]

---

[4]Reprinted by permission from *Concrete Construction Magazine,* World of Concrete Center, 426 South Westgate, Addison, Illinois 60101.

Stating Responsibilities Precisely

Job-related writing, because it is directed to a specific audience, must make absolutely clear what it expects of, or can do for, that audience. Misunderstandings waste time and cost money. Directions on order forms, for example, should indicate how and where information is to be listed and how it is to be routed and acted on. The following directions show readers how to perform a task and/or explain why.

- Enter agency code numbers in the message box.

- Items 1 through 16 of this form should be completed by the injured employee or by someone acting on his or her behalf, whenever an injury is sustained in the performance of duty. The term *injury* includes occupational disease caused by the employment. The form should be given to the employee's official superior within 24 hours following the injury. The official superior is that individual having responsible supervision over the employee.

Other kinds of job-related writing deal with the writer's responsibilities rather than the reader's; for example, "Tomorrow I will meet with the district sales manager to discuss (1) July's sales figures, (2) the necessity of redesigning our website, and (3) next fall's production schedule. I will e-mail a short report of our discussion by August 5, 2009."

Persuading and Offering Recommendations

Persuasion is vital to writing on the job. It is one of the most crucial skills you can learn. It determines how successful you and your company or agency will be. You must establish yourself as an authority on the topic.

Promoting Corporate Image

Much of your writing in the business world is to promote your company's image by persuading customers and clients (a) to buy a product or service or (b) to adopt a plan of action endorsed by your employer. Not only will you have to attract new customers, but you will also have to persuade previous ones to purchase your company's product and/or service again. You will have to convince readers that you (and your company) can save them time and money, increase efficiency, reduce risk, and improve their image.

Expect also to be called upon to write convincingly about your company's image, as in the case of product recalls and discontinuations, customer complaints, or damage control after a corporate mistake affecting the environment.

To be persuasive, you will need to get your readers' attention and communicate clearly so that they understand your message and remember it. A large part of persuasion is supporting claims with evidence. You will have to conduct research; provide logical arguments; supply examples; and, especially important, identify the most relevant information for your audience. Notice how the advertisement in Figure 1.6 offers a bulleted list of persuasive reasons—based on cost, time, safety, efficiency, and convenience—to convince correctional officials that they should use General Medical's services rather than those of a hospital or clinic.

Figure 1.6   An advertisement using arguments based on cost, time, efficiency, safety, and convenience to persuade a potential customer to use a service.

Visual stresses inefficient way to transport prisoners for medical attention

Bulleted list conveniently and persuasively uses factual data to convince

GENERAL MEDICAL WILL STOP THE UNNECESSARY TRANSPORTING OF YOUR INMATES.

- We'll bring our X-ray services to your facility, 7 days a week, 24 hours a day.

    We can reduce your X-ray costs by a minimum of 28%. X-ray cost includes radiologist's interpretation and written report.

    Same day service with immediate results telephoned to your facility.
- Save correctional officers' time, thereby saving your facility money.
- Avoid chance of prisoner's escape and possible danger to the public.
- Avoid long waits in overcrowded hospitals.
- Reduce your insurance liabilities.
- Other Services Available: Ultrasound, Two Dimensional Echocardiogram, C.T. Scan, EKG, Blood Lab and Holter Monitor.

General Medical Is Your On-Site Medical Problem Solver

General Medical Services Corp.

A subsidiary of

Federal Medical Industries, Inc. O.T.C.
950 S.W. 12th Avenue, 2nd Floor Suite, Pompano, Florida 33069
(305) 942-1111   FL WATS: 1-800-654-8282

Writing Persuasively In-House

Writing for individuals you work for and with also requires you to develop your skills at persuading readers. You will be expected to make recommendations to your employer and evaluate various products or options by studying, analyzing, and deciding on the most relevant one(s) for your boss. Your reader will expect you to offer clear-cut, logical, and convincing reasons for your choice.

On the job, you will also be asked to write memos, e-mails, and even letters to boost the morale of employees, encourage them to be more productive, and compliment them on jobs well done. You can also expect to write about (and explain and solve) problems your company faces, such as when a market has shrunk and your boss wants to know what can be done about it or when a service or product your company relies on becomes too costly.

Figure 1.7 (p. 17) contains a persuasive e-mail from an employee to a business manager to report a payroll mistake and to request a reimbursement. Note how the writer provides factual, not impressive, information; attaches a time sheet (a type of visual); gives accurate details; identifies her own and her immediate supervisor's responsibilities; and persuasively, and diplomatically, states her case.

A persuasive e-mail from an employee to a business manager.   **Figure 1.7**

| Reply | Save | Forward | Print | Delete |

To: <lgriffin@starinstruments.com> (Lee Griffin)
From: <rburke@starinstruments.com> (R. Burke)
Date: 19 October 2009 10:05 a.m. EST
Subject: Shortage in October Paycheck

My paycheck for the two-week period ending October 15 was $60.00 short. For this period I should have been paid $540.00. Instead, my check was for only $495.00. I think I know why there may have been a discrepancy. The $60.00 additional pay was the result of my having put in five hours of overtime on October 8 and October 12 (2½ hours each day @ $12.00 per hour). This overtime was not reflected on my current pay stub.

I have double-checked with my supervisor, Gloria Arrelo, who assured me that she recorded my overtime on the timesheets she sent to your office. She has kindly given me a copy that I have scanned and attached to this e-mail to verify my hours.

Thank you for correcting your records and for crediting me with the additional $60.00 for my overtime.

Clearly explains and documents the problem

Offers further evidence

Closes politely with request

Ethical Writing in the Workplace

On-the-job writing involves much more than conveying facts about products, equipment, costs, and the day-to-day operations of a business. Your writing also has to be ethical. Writing ethically means doing what is right and fair and being honest and just with your employer, co-workers, and customers. Your reputation and character plus your employer's corporate image will depend on your following an ethical course of action.

Many of the most significant bywords in the world of business reflect an ethical commitment to honesty and fairness: *accountability, public trust, equal opportunity employer, good faith effort, truth in lending, fair play, honest advertising, full disclosure, high professional standards,* and *community responsibility*.

Unethical business dealings, on the other hand, are captured in *cover-ups, hidden costs, shady deals, spin doctors, accounting fraud, foul play, misrepresentations, price gouging, scams, bias,* and *unfair advantage*. Those are the activities that keep Better Business Bureaus active and make customers angry.

Ten Ethical Requirements on the Job

In the workplace, you will be expected to meet the highest ethical standards by fulfilling the following requirements.

- Supplying honest and up-to-date information about yourself in your résumé and job applications. The résumé (see pp. 136–151) is one key place where most people must make ethical decisions about candor and honesty.
- Respecting co-workers, customers, and vendors in conduct that avoids bullying, discrimination, or any other unfair and unprofessional action.
- Avoiding language that excludes others on the basis of gender, race, national origin, age, or physical condition (see pp. 43–46).
- Maintaining accurate and current records at work. Remember: "If it isn't written, it didn't happen."
- Complying with all local, state, and federal regulations, especially those ensuring a safe, healthy work environment, products, and/or service; for example, those of the Occupational Safety and Health Administration (OSHA).
- Adhering to your profession's code or standard of ethics, internal audits, licenses, and certificate requirements.
- Following your company's policies and procedures.
- Honoring guarantees and warranties and meeting customer needs impartially.
- Cooperating fairly with your collaborative team.
- Respecting all copyright obligations and privileges.

Following these guidelines is not only an ethical requirement, but it could also be a legal one. For example, doing personal (or outside consulting) work on company time, padding expense accounts, using company equipment for personal use, or accepting a bribe is unethical and illegal. It would also be neglectful and unethical to allow an unsafe product to stay on the market just to spare your company the expense of a recall. It would be wrong, legally and ethically, to share information about your employer's patent plans on the Internet.

Computer ethics, especially when using the Internet, are essential in the world of work. It would be grossly unethical to erase a computer program intentionally, to violate a software licensing agreement, or to misrepresent (by fabrication or exaggeration) the scope of a database. Follow the "Ten Commandments of Computer Ethics" prepared by the Computer Ethics Institute and listed in Figure 1.8. (See also "Guidelines for Using E-Mail" in Chapter 2, pp. 72–75.)

Writing for the world of multinational corporations places additional ethical demands on you as a writer. You have to make sure you respect the ethics of the foreign countries where your firm does business. Some behavior regarded as normal or routine in the United States might be seen as highly unethical elsewhere. And you should be on your ethical guard not to take advantage of a host country, such as using pesticides or conducting experiments outlawed in the United States.

Some Guidelines to Help You Reach Ethical Decisions

The workplace presents all sorts of conflicts over who is right and who is wrong, what is best for the company and what is not, and whether a service or product should be

The Ten Commandments of Computer Ethics.   **Figure 1.8**

1. Thou shalt not use a computer to harm other people.
2. Thou shalt not interfere with other people's computer work.
3. Thou shalt not snoop around in other people's computer files.
4. Thou shalt not use a computer to steal.
5. Thou shalt not use a computer to bear false witness.
6. Thou shalt not copy or use proprietary software for which you have not paid.
7. Thou shalt not use other people's computer resources without authorization or proper compensation.
8. Thou shalt not appropriate other people's intellectual output.
9. Thou shalt think about the social consequences of the program you are writing or the system you are designing.
10. Thou shalt always use a computer in ways that ensure consideration and respect for your fellow humans.

Computer Ethics Institute, London.

changed and why. You will be asked to take a stand. Here are a few guidelines to help you respond ethically.

1. Follow your conscience and "to thine own self be true." You cannot authorize something that you believe is wrong, dangerous, unfair, contradictory, or incomplete. But don't be hasty. Leave plenty of room for diplomacy and for careful questioning. Don't blow a small matter out of proportion.

2. Be suspicious of convenient (and false) appeals that go against your beliefs. Watch out for the red flags that anyone places in the way of your conscience: "No one will ever know." "It's OK to cut corners every once in a while." "We got away with it last time." "Don't rock the boat." "No one's looking." "As long as the company makes money, who cares?" These rationalizations are traps you must avoid.

3. Maintain good faith in meeting your obligations to your employer, your co-workers, your customers, and your community. It is unethical to lie, exaggerate, or even dodge an issue. Keeping information from a co-worker who needs it, using a password belonging to someone else, omitting a fact, justifying unnecessary expenses—all are unethical acts, just as cheating on an examination or plagiarizing are unethical in your schoolwork.

4. Take responsibility for your actions. Saying "I do not know" when you do know can constitute a serious ethical violation. Keep up-to-date and accurate records. Sign and date your work. Never backdate a document to delete information or to fix an error that you committed. Such action constitutes a cover-up and a serious breach of ethics.

5. Document your work carefully and honestly. Rely on hard evidence: documentation, tests, testimony, valid precedents. Do your homework by studying code books and agency handbooks; confer with a customer or a co-worker when you are in doubt about a major issue. Familiarize yourself with your company's protocols, methods, and materials. Make sure your documents are accurate and comply with appropriate city, state, federal, and international regulations.

6. Weigh all sides before you commit to a conclusion. Research what you write and communicate orally. Rely on hard evidence: documentation, testimony, valid precedents. Do your homework by studying code books and agency handbooks; confer with a customer or a co-worker when you are in doubt about a major issue. Familiarize yourself with company protocols and specifications about the procedures, methods, and materials of your job.

Ethical Dilemmas

Sometimes in the workplace you will face situations where there is no clear-cut right or wrong choice, even with regard to the six ethical categories described above. Here are a few scenarios, similar to ones in which you may find yourself, that are gray areas, ethically speaking, along with some possible solutions.

- You see an opening for a job in your area but the employer wants someone with a minimum of two years of field experience. You have just completed an internship and had one summer's (12 weeks) experience, which together total almost 5 months. Should you apply for the job, describing yourself as "experienced"?

  Yes, but honestly state the type and the extent of your field experience and the conditions under which you obtained it.

- You work for a company that usually assigns commissions to the salesperson for whom a customer asks. A customer asks for a salesperson who happens to have the day off. You assist the customer all afternoon, and even arrange to have an item shipped overnight so that the customer can have it in the morning. When you ring up the sale, should you list your employee number for the commission or the off-duty employee's number?

  You probably should defer crediting the sale to either number until you speak to the absent employee and suggest a compromise—splitting the commission.

- A piece of computer equipment, scheduled for delivery to your customer the next day, arrives with a damaged part. You replace it before the customer receives it. Should you inform the customer?

  Yes, but assure the customer that the equipment is still under the same warranty and that the replacement part is new and also under the same warranty. If the customer protests, agree to let him or her use the computer until a new unit arrives.

As these brief scenarios suggest, sometimes you have to make concessions and compromises to be ethical in the world of work. But other times you just can't; as Figure 1.8 states.

Writing Ethically

Your writing as well as your behavior must be ethical. Words, like actions, have implications and consequences. If you slant your words to conceal the truth or to gain an unfair advantage, you are not being ethical. False advertising is false writing. In your written work, strive to be fair, reliable, and accurate, in reporting events and figures honestly and without bias or omissions.

Unethical writing is usually guilty of one or more of the following faults, which can conveniently be listed as the three **M**'s: misquotation, misrepresentation, and manipulation. Here are eight examples.

1. Plagiarism is stealing someone else's words (work) and claiming it as your own. At work, plagiarism is unethically claiming a co-worker's ideas, input, or report as your contribution. In a research report or paper, you are guilty of plagiarism if you use another person's words (or even a rough paraphrase) without documenting the source. Do not think that by changing a few words here and there you are not plagiarizing. Copying someone else's software or downloading their visuals are also an act of plagiarism. Give proper credit to your source, whether in print, in person (through an interview), or online.

The penalties for plagiarism are severe—a reprimand or even the loss of your job. At school, you run the risk of failing the course or, worse, being expelled. See pages 284–285 for advice on how to avoid plagiarism.

2. Selective misquoting deliberately omits damaging or unflattering comments to paint a better (but untruthful) picture of you or your company. By picking and choosing words from a quotation, you unethically misrepresent what the speaker or writer originally intended.

> Full Quotation: I've enjoyed at times our firm's association with Technology, Inc., although I was troubled by the uneven quality of their service. At times, it was excellent while at others it was far less so.
>
> Selective Misquotation: I've enjoyed . . . our firm's association with Technology, Inc. The quality of their service was . . . excellent.

The dots, called *ellipses,* unethically suggest that only extraneous or unimportant details were omitted. See page 342 on how to use ellipses properly.

3. Arbitrary embellishment of numbers unethically misrepresents, by increasing or decreasing percentages or other numbers, statistical or other information. It is unethical to stretch the differences between competing plans or proposals to gain an unfair advantage or to express accurate figures in an inaccurate way.

> Embellishment: Our competitor's sales volume increased by only 10 percent in the preceding year while ours doubled.

> Ethical: Our competitor controls 90 percent of the market, yet we increased our share of that market from 5 percent to 10 percent last year.

4. Manipulation of data or context, closely related to #3, is the misrepresentation of events, usually to "put a good face" on a bad situation. The writer here unethically uses slanted language and intentionally misleading euphemisms to misinterpret events for readers.

> Manipulation: Looking ahead to 2009, the United Funds Group is exceptionally optimistic about its long-term prospects in an expanding global market. We are happy to report steady to moderate activity in an expanding sales environment last year. The United Funds Group seeks to build on sustaining investment opportunities beneficial to all subscribers.
>
> Ethical: Looking ahead to 2009, the United Funds Group is optimistic about its long-term prospects in an expanding global market. Though the market suffered from inflation this year, the United Funds Group hopes to recoup its losses in the year ahead.

The writer minimizes the negative effects of inflation by calling it "an expanding sales environment."

5. Using fictitious benefits to promote a product or service that seemingly promises customers advantages but delivers none.

> False Benefit: Our bottled water is naturally hydrogenated from clear underground springs.
>
> Truth: All water is hydrogenated because it contains hydrogen.

6. Unfairly characterizing (by exaggerating or minimizing) hiring or firing conditions is unethical.

> Unethical: Our corporate restructuring will create a more efficient and streamlined company, benefiting management and workers alike.
>
> Truth: Downsizing has led to 150 layoffs.

7. Manipulating international readers by adopting a condescending view of their culture and economy is unethical.

> Unethical: Since our product has appealed to U.S. customers for the last sixteen months, there's no doubt that it will be popular in your country as well.
>
> Fair: Please let us know if any changes in design or construction may be necessary for customers in your country.

8. Misrepresenting through distorting or slanted visuals is one of the most common types of unethical communication. Making a product look bigger, better, or more professional is all too easy with graphics software packages. Making warning or caution statements the same size and type font as ingredients or directions or enlarging advertising hype (Double Your Money Back) is unethical if major points are reduced to small print.

Because ethics is such an important topic in writing for the workplace, this key idea will be stressed throughout this book. (See, for example, Chapter 3, p. 72, and Chapter 8, p. 241.)

Revision Checklist

At the end of each chapter is a checklist you should review before you submit the final copy of your work, either to your instructor or to your boss. Regard each checklist as a summary of the main ideas in the chapter as well as a handy guide to quality control. You may find it helpful to check each box as you verify that you have performed the necessary revision and review. Effective writers are also careful editors.

☐ Identified my audience—their background, knowledge of English, reason for reading my work, and likely response to my work and me.

☐ Showed respect for and appropriately shaped my message for a global marketplace.

☐ Made it clear what I want my audience to do after reading my work.

☐ Tailored the message to my audience's needs and background, giving them neither too little nor too much information.

☐ Pushed to the main point right away; did not waste my reader's time.

☐ Selected the most appropriate language, technical level, tone, and level of formality.

☐ Did not waste my audience's time with unsupported generalizations or opinions; instead gave them accurate measurements, facts, and carefully researched material.

☐ Included appropriate visuals for my audience.

☐ Used persuasive reasons and data to convince readers to accept my plan or work.

☐ Ensured that my writing is ethical—accurate, complete, honest, a true reflection of the situation or condition I am explaining or describing.

☐ Followed the "Ten Commandments of Computer Ethics" (see Figure 1.8).

☐ Gave full credit to sources I used, including resource people.

☐ Avoided plagiarism and unfair or dishonest use of copyrighted materials, both written and visual, including all electronic media.

Exercises

1. Write a memo (see Chapter 3, pp. 65–66 for format) addressed to a prospective supervisor to introduce yourself. Your memo should have four headings: **education**—including goals and accomplishments; **job information**—where you have worked and your responsibilities; **community service**—volunteer work, church work, youth groups; and **writing experience**—your strengths and what you would like to see improved.

Additional Activities related to writing and your career are located at **college.cengage .com/pic/ kolinconcise2e.**

2. Bring to class a set of printed instructions from a memo, a sales letter, or a brochure. Comment on how well the printed material answers the following questions.
   a. Who is the audience?
   b. Why was the material written?
   c. What is the message?
   d. Are the style and tone appropriate for the audience, the purpose, and the message? Why?
   e. Discuss the use of color in the document. How does color (or the lack of it) affect an audience's response to the message?

3. Cut out a newspaper ad that contains a drawing or photograph. Bring it to class together with a paragraph of your own (75–100 words) describing how the message of the ad is directed to a particular audience and commenting on why the illustration was selected for that audience.

4. Pick one of the following topics and write two descriptions of it. In the first description, use technical vocabulary; in the second, use language suitable for the general public.
   a. spark plug
   b. blood pressure cuff
   c. carburetor
   d. computer chip
   e. camera
   f. legal contract
   g. satellite radio
   h. cyberspace
   i. iPod
   j. protein
   k. high-definition TV
   l. DVD
   m. bread
   n. money
   o. color scanner
   p. soap
   q. blogging
   r. X-Box game
   s. AIDS
   t. thermostat
   u. trees
   v. earthquake
   w. CD burner
   x. euro

5. Select one article from a newspaper and one article from either a professional journal in your major field or one of the following journals: *American Journal of Nursing, Business Marketing, Business Week, Computer, Construction Equipment, Criminal Justice Review, E-Commerce, E—The Environmental Magazine, Food Service Marketing, Global Investor, Journal of Forestry, Latino Future, National Safety News, Nutrition Action, Office Machines, Park Maintenance, Scientific American, Today's Black Woman, Women International.* State how the two articles you selected differ in terms of audience, purpose, message, style, and tone.

6. Assume that you work for Appliance Rentals, Inc., a company that rents TVs, microwave ovens, home theater systems, and the like. Write a persuasive letter to the members of a campus organization or civic club urging them to rent an appropriate appliance or appliances. Include details in your letter that might have special relevance to members of this specific organization.

7. How do the visuals and the text of the Sodexho advertisement on page 25 stress to current (and potential) employees, customers, and stockholders that the company is committed to human diversity in the workplace? How would the ad appeal to international readers? Also explain how the ad illustrates the functions of on-the-job writing defined on pages 12–16, especially persuasion.

8. Write a letter to an Internet service provider that has mistakenly billed you for caller ID equipment you never ordered, received, or needed.

"*I am* making
a difference."

"*I am*
improving
your life."

"*I am* taking care
of you. And people
you care about."

People define our success.

Diverse perspectives and

talents allow us to provide

innovative food and

I am
Sodexho

facilities management

services that improve the

quality of daily life for

the millions of people we

serve in the U.S. every day.

"*I am* ensuring
your safety."

"*I am* a step ahead."

Committed to Diversity and Inclusion

★ ★
★ ★
Sodexho

sodexhoUSA.com • 1-800-SODEXHO

©Sodexho *Member of Sodexho Alliance®*

Food Services, Facilities Management, Vending, Catering, Office Refreshment
Services, Environmental Services, Landscaping & Grounds Management,
Conferencing, Plant Operations & Management

An advertisement for Sodexho.

9. The following statements contain embellishments, selected misquotations, false benefits, and other types of unethical tactics. Revise each statement to eliminate the unethical aspects.

   a. Storm damage done to water filtration plant #3 was minimal. While we had to shut down temporarily, service resumed to meet residents' needs.

   b. All customers qualify for the maximum discount available.

   c. The service contract . . . on the whole . . . applied to upgrades.

   d. We followed the protocols precisely with test results yielding further opportunities for experimentation.

   e. All of our costs were within fair-use guidelines.

   f. Customers' complaints have been held to a minimum.

   g. All of the lots we are selling offer easy access to the lake.

10. Your company is regulated and inspected by the Environmental Protection Agency. In ninety days, the EPA will relax a particular regulation about dumping industrial waste. Your company's management is considering cutting costs by relaxing the standard now, before the new, easier regulation is in place. You know that the EPA inspector probably will not return before the ninety-day period elapses. What do you recommend to management?

CHAPTER 2

The Writing Process and Collaboration at Work

In Chapter 1 you learned about the different functions of writing for the world of work and also explored some basic concepts all writers must master. To be a successful writer, you need to

- identify your audience's needs
- determine your purpose in writing to that audience
- make sure your message meets your audience's needs
- use the most appropriate style and tone for your message
- format your work to convey your message clearly for your audience

Just as significant to your success is knowing how effective writers actually create their work for their audiences. This chapter gives you practical information about the strategies and techniques careful writers use when they work. These procedures are a vital part of what is known as the *writing process*. This process involves matters such as how writers gather information, how they transform their ideas into written form, and how they organize, revise, and edit what they have written to ensure that it is suitable for their readers.

To expand your understanding of the writing process and collaborative writing, take advantage of the Web Links, Additional Activities, and ACE Self-Tests at **college.cengage .com/pic/ kolinconcise2e.**

What Writing Is Not and Is

As you begin your study of writing for the world of work, it might be helpful to identify some notions about what writing is and what it is not.

What Writing Is Not

- **Writing is not something mysterious done according to a magical formula known only to a few.** Even if you have not done much writing before, you can learn to write effectively.
- **Writing is not simply a hit-or-miss affair, left up to chance.** Successful writing requires hard work and thoughtful effort. It is not done well by simply going through an ordered set of steps as if you were painting by number. You cannot

sit down for fifteen minutes and expect to write the perfect memo, letter, or short report straight through. Writing does not proceed in some predictable way, in which introductions are always written first and conclusions last.

- **Just because you put something on paper or online does not mean it is permanent and unchangeable.** Writing means *re*writing, *re*vising, *re*thinking. The better a piece of writing is, the more the writer has reworked it.

What Writing Is

- **Writing is a fluid process**—it is dynamic, not static. It enables you to discover and evaluate your thoughts as you draft and revise.
- **A piece of writing changes as your thoughts and information change** and as your view of the material changes.
- **Writing means making a number of judgment calls.**
- **Writing takes time.** Some people think that revising and polishing are too time consuming. But poor writing actually takes more time and costs more money in the end. It can lead to misunderstandings, lost sales, product recalls, and even damage to your reputation and that of your company.
- **Writing grows sometimes in bits and pieces and sometimes in great spurts.** It needs many revisions; an early draft is never a final copy.

▌ Researching

Before you start to compose any e-mail, memo, letter, or report, you'll need to do some research. Research is crucial to obtain the right information for your audience. Information must be factually correct and intellectually significant. The world of work is based on conveying information—the logical presentation and sensible interpretation of facts.

Don't ever think you are wasting time by not starting to write your report or letter immediately. Actually, you will waste time and risk doing a poor job if you do not find out as much as possible about your topic (and your audience's interest in it).

First, find out as much as you can about the nature of your assignment and your readers, and what they expect from your written work. Next, determine the exact kind of research you must do to gather and interpret the information your audience needs. Your research can include

- interviewing people inside and outside your company
- doing fieldwork or performing lab studies
- preparing for conferences to ask appropriate questions
- collaborating with colleagues and supervisors in person or by e-mail
- distributing a questionnaire to conduct a survey
- reading current periodicals, reports, and other documents
- evaluating reports, products, services, and websites
- getting briefings from sales or technical staff
- contacting vendors, customers, and inspectors

Keep in mind that research is not confined to just the beginning of the writing process; it goes on throughout.

Planning

At this stage in the writing process, your goal is to get something—anything—down on paper or on your computer screen. For most writers, getting started is the hardest part of the job. But you will feel more comfortable and confident once you begin to see your ideas before your eyes. It is always easier to clarify and criticize something you can see.

Getting started is also easier if you have researched your topic, because you have something to say and to build on. Each part of the process relates to and supports the next. Careful research prepares you to begin writing.

Still, getting started is not easy. Take advantage of a number of widely used strategies that can help you to develop, organize, and tailor the right information for your audience. Use any one of the following techniques, alone or in combination.

   1. Clustering. In the middle of a sheet of paper, write the word or phrase that best describes your topic, then start writing other words or phrases that come to mind. As you write, circle each word or phrase and connect it to the word from which it sprang. Note the clustered grouping in Figure 2.1 for a report encouraging a manager to switch to flextime—a system in which employees can work on a flexible time schedule within certain limits. The resulting diagram gives the writer a rough sense of some of the major divisions of the topic and where they may belong in the report.

   2. Brainstorming. At the top of a sheet of paper or your computer screen, describe your topic in a word or phrase and then list any information you know or found out about that topic—in any order and as quickly as you can. Brainstorming is like thinking aloud except that you are recording your thoughts.

   - Don't stop to delete, rearrange, or rewrite anything, and don't dwell on any one item.
   - Don't worry about spelling, punctuation, grammar, or whether you are using words and phrases instead of complete sentences.
   - Keep the ideas flowing. The result may well be an odd assortment of details, comments, and opinions.
   - After stepping away from the list for a few minutes or an hour or two and returning with fresh eyes, add and delete some ideas or combine or rearrange others as you start to develop your topic in more detail.

   Figure 2.2 (p. 31) shows Marcus Weekley's initial brainstormed list for a report to his boss on purchasing a new color laser printer. After he began to revise it, he realized that some items were not relevant for his audience (6, 8, and 13). He also recognized that some items were repetitious (1, 2, and 11). Further investigation revealed that his company could purchase a printer for far less than his initial high guess (17). As Weekley continued to work on his list and the overall topic became clearer to him, he added and deleted points.

Figure 2.1   Clustering on the topic of flextime.

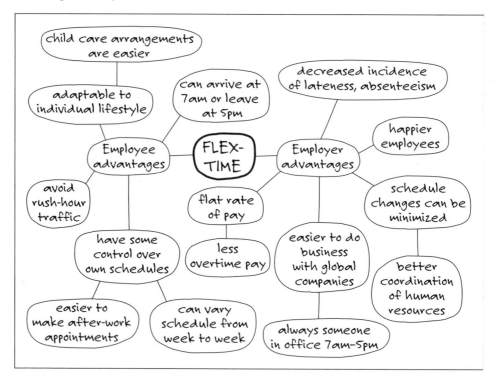

3. Outlining. For most writers, outlining may be the easiest and most comfortable way to begin or to continue planning their report or letter. Outlines typically go through several revisions, and are for no one's eyes but your own, so don't worry if your first attempt is brief, messy, informal, or incomplete. Clarify and tighten your outline as you revise. Note how Marcus Weekley organized his revised brainstormed list into an outline (in Figure 2.3, p. 32).

▮ Drafting

If you have done your planning carefully, you will find it easier to start your first draft. When you draft, you convert the words and phrases from your outlines, brainstormed lists, or clustered groups into paragraphs. During drafting, as elsewhere in the writing process, you will see some overlap as you look back over your lists or outlines to shape your text.

Don't expect to wind up with a polished, complete version of your paper after working on only one draft. In most cases, you will have to work through many drafts, but each draft should be less rough and more acceptable than the preceding one.

Marcus Weekley's initial, unrevised brainstormed list.   **Figure 2.2**

1. combines four separate pieces of equip—printer, copier, fax, and scanner

2. more comprehensive than our current configuration of four pieces of equip

3. would coordinate with office furniture

4. energy efficiency increased due to fewer machines being used

5. more scalable fonts

6. one machine interfaces with all others in same-case housing

7. scanner makes photographic-quality pictures

8. print capabilities are a real contribution to technology

9. increase communication abilities through fax machine

10. new scanner picture quality better than current scanner

11. only have to buy one machine as opposed to four

12. speed of fax allows quick response time

13. stock is doing better on Wall Street compared to other stocks

14. reducing our advertising costs through use of color printer

15. increase work area available

16. would help us do our work better

17. top of line models can be bought for $4,500

List is not organized but simply reflects writer's ideas about possible topics

1, 2, and 11 are repetitious

6, 8, and 13 are not relevant

Research will show price is too high

Figure 2.3   Marcus Weekley's early outline after revising his brainstormed list.

Outline form helps writer group ideas/ topics and go to next step in revision

Outline headings correspond to major sections of report

Last section of outline also functions as a conclusion

I. Convenience/Capabilities

A. Would reduce number of machines required to service

B. Can be configured easily for our network system

C. Easy to install and operate

D. 35.6 Kbps fax machine would increase our communication

E. 2,400 × 1,200 dpi copier means better copy quality

F. 4,800 × 4,800 dpi scanner means higher quality pictures than current scanner

II. Time/Efficiency

A. 50 ppm printer is nearly twice as fast as current printer

B. 35.6 Kbps speed fax allows quick response time

C. Greater graphics capability—90 scalable fonts

D. Scanner compatible with our current PhotoEdit imaging/graphics software

E. 100,000-page monthly duty cycle means less maintenance

III. Money

A. Costs less overall for multitasking printer than combined four machines

B. Reduced monthly power bill by using one machine rather than four

C. Included ScanText and WordPort software means not buying new software to network office computers

D. Save on service costs

E. Reduced advertising costs through printer 50–1,000% enlargement/ reduction options, which allow for more in-house advertising

Key Questions to Ask as You Draft

As you work on your drafts, ask yourself the following questions about your content and organization.

- Am I giving my readers too much or too little information?
- Does this point belong where I have it, or would it more logically follow or precede something else?
- Is this point necessary and relevant?
- Am I repeating or contradicting myself?
- Have I ended appropriately for my audience?

To answer the questions successfully, you may have to continue researching your topic and reexamining your audience's needs. But, during the process, new and even better ideas may come to you.

Guidelines for Successful Drafting

The following are some suggestions to help your drafting go more smoothly and efficiently.

- In an early draft, write the easiest part first. Some writers feel more comfortable drafting the body (or middle) of their work first. See the differences between the memo in Figures 2.4 and 2.5, pages 35–36.
- As you work on a later draft, write straight through. Do not worry about spelling, punctuation, or the way a word or sentence sounds. Save those concerns for later stages.
- Allow enough time between drafts so that you can evaluate your work with fresh eyes and a clear mind.
- Get frequent outside opinions. Show, e-mail, or fax a draft to a fellow student, a co-worker, or maybe a supervisor for comment. A new pair of eyes will see things you missed. Collaboration is essential in the workplace (see pp. 46–47).
- Start considering if visuals, and what kind, would enhance the quality of your work and where they might best be positioned.

▌Revising

Revision is an essential stage in the writing process. It requires more than giving your work another quick glance. Do not be tempted to skip the revision stage just because you have written the required number of words or sections or because you think you have put in too much time already. Revision is done *after* you produce a draft that you think conveys the appropriate message for your audience. The quality of your memo, letter, or report depends on the revisions you make now.

Allow Enough Time to Revise

Like planning or drafting, revision is not done well in one big push. It evolves over a period of time. Make sure you budget enough time to do it carefully.

- Avoid drafting and revising in one sitting. If possible, wait at least a day before you start to revise. (In the busy work world, waiting a couple of hours may have to suffice.)
- Ask a co-worker or friend familiar with your topic to comment on your work.
- Plan to read your revised work more than once.

Revision Is Rethinking

When you revise, you *resee*, *rethink*, and *reconsider* your entire document. You ask questions about the major issues of content, organization, and tone. Revision involves going back and repeating earlier steps in the writing process.

Revision means asking again the questions you have already asked and answered during the planning and drafting stages. During the process, you will discover gaps to fill, points to change, and errors to correct in your draft. Revision gives you a second (or third or fourth) chance to get things right for your audience.

Key Questions to Ask as You Revise

Content

1. Is it accurate? Are my facts (figures, names, addresses, dates, costs, references, warranty terms, statistics) correct?
2. Is it relevant for my audience and purpose? Have I included information that is unnecessary, too technical, inappropriate?
3. Have I given enough evidence to explain things adequately and to persuade my readers? (Too little information will make readers skeptical about what you are describing or proposing.) Have I left anything out?

Organization

1. Have I clearly identified my main points and shown readers why those points are important?
2. Is everything in the right, most effective order? Should anything be switched or moved closer to the beginning or the end of my document?
3. Have I spent too much (or too little) effort on one section? Do I repeat myself? What can be cut?
4. Have I grouped related items in the same part of my report or letter, or have I scattered details that really need to appear together in one paragraph or section?

Tone

1. How do I sound to my readers—professional and sincere, or arrogant and unreliable? What attitude do my words or expressions convey?
2. How will my readers think I perceive them—honest and intelligent or unprofessional and uncooperative?

A "Before" and "After" Revision

Figures 2.4 and 2.5 show a "before" and "after" revision. The writer, Mary Fonseca, a staff member at Seacoast Labs, was asked by her supervisor to prepare a short report for the general public on the lab's most recent experiments. One of her later drafts begins with the paragraphs in Figure 2.4.

When starting to revise Figure 2.4, Fonseca realized that it lacked focus. It jumped back and forth between *drag* on ships and on an airplane. Since Seacoast Labs did not work on planes, she wisely decided to drop that idea. She also realized that the information on the effects of drag, something Seacoast was working on, was so important it deserved a separate paragraph. In light of this key idea, she realized that her explanation of molecules, eddies, and drag needed to be made more reader-friendly. Researching further, she decided to add a new paragraph on the causes and effects of drag, which became paragraph 2 in Figure 2.5.

Yet by pulling ideas about drag and its effects from her longish first paragraph in Figure 2.4, Fonseca was left with the job of finding an opening for her report. Buried in her original opening paragraph was the idea that we cannot always see the forces of nature but "we can feel them." She thought this analogy of walking against the wind and drag would work better for her audience of nonspecialists than the original wooden remarks she had started with.

Opening, unorganized paragraphs of Mary Fonseca's draft.   Figure 2.4

Drag is an important concept in the world of science and technology. It has many implications. Drag occurs when a ship moves through the water and eddies build up. Ships on the high seas have to fight the eddies, which results in drag. In the same way, an airplane has to fight the winds at various altitudes at which it flies; these winds are very forceful, moving at many knots per hour. All these forces of nature are around us. Sometimes we can feel them, too. We get tired walking against a strong wind. The eddies around a ship are the same thing. These eddies form various barriers around the ship's hull. They come from a combination of different molecules around the ship's hull and exert quite a force. Both types of molecules pull against the ship. This is where the eddies come in.

Information is hard to follow and not relevant for audience

Scientists at Seacoast Labs are concerned about drag. Dr. Karen Runnels, who joined Seacoast about three years ago, is the chief investigator. She and her team of highly qualified experts have constructed some fascinating multilevel water tunnels. These tunnels should be useful to ship owners. Drag wastes a ship's fuel.

Important point not developed

Figure 2.5   A revision of Mary Fonseca's draft in Figure 2.4.

Effective use of definition and headings

Describes cause and effect of drag

Uses an easy-to-follow analogy

Clearly explains the research and its importance for readers

What is drag?

We cannot see or hear many of the forces around us, but we can certainly detect their presence. Walking or running into a strong wind, for example, requires a great deal of effort and often quickly leaves us feeling tired. When a ship sails through the water, it also experiences these opposing forces known as **drag**. Overcoming drag causes a ship to reduce its energy efficiency, which leads to higher fuel costs.

How drag works

It is not easy for a ship to fight drag. As the ship moves through the water, it drags the water molecules around its hull at the same rate the ship is moving. Because of the cohesive force of those molecules, other water molecules immediately outside the ship's path get pulled into its way. All the molecules become tangled rather than simply sliding past each other. The result is an eddy, or small circling burst of water around the ship's hull, which intensifies the drag. Dr. Charles Hester, a noted engineer, explains it using an analogy: "When you put a spoon in honey and pull it out, half the honey comes out with the spoon. That's what is happening to ships. The ship is moving and at the same time dragging the ocean with it."

At Seacoast Labs, scientists are working to find ways to reduce drag on ships. Dr. Karen Runnels, the principal investigator, and a team of researchers have constructed water tunnels to simulate the movement of ships at sea. The drag a ship encounters is measured from the tiny air bubbles emitted in the water tunnel. Dr. Runnels's team has also developed the use of polymers, or long carbon chain molecules, to reduce drag. The polymers act like a slimy coating for the ship's hull to help it glide through the water more easily. When asbestos fibers were added to the polymer solutions, the investigators measured a 90 percent reduction in drag. The team has also experimented with an external pump attached to the hull of a ship, which pushes the water away from a ship's path.

Although her original draft organization and ideas were now far better than those in Figure 2.4, she realized she had said very little about her employer Seacoast Labs. Doing more research, she found information about Seacoast's experiments and why they were so important in saving money. This information was far more important and relevant than saying Dr. Runnels had been at Seacoast for three years.

Through revision and further research, then, Mary Fonseca transformed two poorly organized and incomplete paragraphs into three separate yet logically connected ones that highlighted her employer's work. Thanks to her revision, too, she came up with two very helpful headings—"What is drag?" and "How drag works"—for her nonspecialist readers (see Figure 2.5).

Editing

Editing is quality control for your reader. This last stage in the writing process might be compared to detailing an automobile—the preparation a dealer goes through to ready a new car for prospective buyers. Editing is done only after you are completely satisfied that you have made all the big decisions about content and organization—that you have said what you wanted to, where and how you intended, for your audience.

When you edit, you will check your work for

- sentences
- word choices
- punctuation / spelling
- grammar and usage
- clarity

As with revising, don't skip or rush through the editing process, thinking that once your ideas are down, your work is done. Your style, punctuation, spelling, and grammar matter a great deal to readers. If your work is hard to read or contains mistakes in spelling or punctuation, readers will think that your ideas and your research are also faulty. Also edit for sexist language (see pp. 43–45).

The following sections give you some basic guidelines about what to look for when you edit your sentences and words. The appendix, "A Writer's Brief Guide to Paragraphs, Sentences, and Words" (pp. 329–346), contains helpful suggestions on correct spelling and punctuation.

Guidelines for Writing Lean and Clear Sentences

The most frequent complaints readers voice about poorly edited writing in the world of work are:

- The sentences are too long. I could not follow the writer's meaning.
- The sentences are too complex. I could not understand what the writer meant the first time I read the work; I had to reread it several times.

- The sentences are unclear. Even after I reread them, I am not sure I understood the writer's message.
- The sentences are too short and simplistic. The writing seemed immature.

Wordy, unclear sentences frustrate readers by forcing them to reread. Short, simplistic sentences also frustrate readers with their lack of sophistication. Writing clear, readable sentences is not always easy. It takes effort, but the time you spend editing will pay off in rich dividends for you and your readers. The guidelines that follow should help with the editing phase of your work.

1. Avoid Needlessly Complex or Lengthy Sentences

Do not pile words on top of words. Instead, edit one overly long sentence into two or even three more manageable ones.

> **Too Long:** The planning committee decided that the awards banquet should be held on May 15 at 6:30, since the other two dates (May 7 and May 22) suggested by the hospitality committee conflict with local sports events, even though one of those events could be changed to fit our planning needs.
>
> **Edited for Easier Reading:** The planning committee has decided to hold the awards banquet on May 15 at 6:30. The other dates suggested by the hospitality committee—May 7 and May 22—conflict with two local sports events.

2. Combine Short, Choppy Sentences

Don't shorten long, complex sentences only to turn them into choppy, simplistic ones. A memo, e-mail, or letter written exclusively in short, staccato sentences sounds immature.

When you find yourself looking at a series of short, blunt sentences, as in the following example, combine them where possible and use connective words similar to those italicized in the edited version.

> **Choppy:** Medical transcriptionists have many responsibilities. Their responsibilities are important. They must be familiar with medical terminology. They must listen to dictation. Sometimes physicians talk very fast. Then the transcriptionist must be quick to transcribe what is heard. Words could be missed. Transcriptionists must forward reports. These reports have to be approved. This will take a great deal of time and concentration. These final reports are copied and stored properly for reference.
>
> **Edited:** Medical transcriptionists have many important responsibilities. *These* include transcribing physicians' orders accurately and using correct medical terminology. *When* physicians talk rapidly, transcriptionists have to make sure that no words are omitted. Also *among the most demanding* of their duties are forwarding transcriptions *and then*, after approval, storing copies properly for future reference.

3. Edit Sentences to Tell Who Does What to Whom or What

The clearest sentence structure in English is the subject-verb-object (s-v-o) pattern.

    s   v   o

Sue mowed the grass.

    s   v   o

Our website contains a link to key training software programs.

Readers find this pattern easiest to understand because it provides direct and specific information about the action. Hard-to-read sentences obscure or scramble information about the subject, the verb, or the object. In the following unedited sentence, subjects are hidden in the middle rather than being placed in the most crucial subject position.

> **Unclear:** The control of the ceiling limits of glycidyl ethers on the part of the employers for the optimum safety of workers in the workplace is necessary. (Who is responsible for taking action? What action must they take? For whom is such action taken?)
>
> **Edited:** Employers must control the ceiling limits of glycidyl ethers for the workers' safety.

4. Use Strong, Active Verbs Rather Than Verb Phrases

In trying to sound important, bureaucratic writers often avoid using simple, graphic verbs. Instead, these writers use a weak verb phrase (for example, *provide mainte-nance of* instead of *maintain, work in cooperation with* instead of *cooperate*). Such verb phrases imprison the active verb inside a noun format and slow a reader down. Note how the edited version here rewrites the weak verb phrase.

> **Weak:** The city provided the employment of two work crews to assist the strengthening of the dam.
>
> **Strong:** The city employed two work crews to strengthen the dam.

5. Avoid Piling Modifiers in Front of Nouns

Putting too many modifiers (words used as adjectives) in the reader's path to the noun will confuse the reader, who cannot decipher how one modifier relates to another modifier or to the noun. To avoid that problem, edit the sentence to place some of the modifiers after or before the nouns they modify.

> **Crowded:** The vibration noise control heat pump condenser quieter can make your customer happier.
>
> **Readable:** The quieter on the condenser for the heat pump will make your customer happier by controlling noise and vibrations.

6. Replace Wordy Phrases or Clauses with One- or Two-Word Synonyms

> **Wordy:** The college has parking zones for different areas for people living on campus as well as for those who do not live on campus and who commute to school.
>
> **Edited:** The college has different parking zones for resident and commuter students. (Twenty words of the original sentence—everything after "areas for"—have been reduced to four words: "resident and commuter students.")

7. Combine Sentences Beginning with the Same Subject or Ending with an Object That Becomes the Subject of the Next Sentence

Wordy: I asked the inspector if she were going to visit the plant this afternoon. I also asked her if she would come alone.

Edited: I asked the inspector if she were going to visit the plant alone this afternoon.

Wordy: Homeowners want to buy low-maintenance bushes. These low-maintenance bushes include the ever-popular holly and boxwood varieties. These bushes are also inexpensive.

Edited: Homeowners want to buy low-maintenance and inexpensive bushes such as holly and boxwood. (This revision combines three sentences into one, condenses twenty-four words into fourteen, and joins three related thoughts.)

Guidelines for Cutting Out Unnecessary Words

Too many people in business think the more words, the better. Nothing could be more self-defeating. Your readers are busy; unnecessary words slow them down. Make every word work. Cut out any words you can from your sentences. If the sentence still makes sense and reads correctly, you have eliminated wordiness. For example, the phrases on the left should be replaced with the precise words on the right.

Wordy	Concisely Edited
at a slow rate	slowly
at this point in time	now
be in agreement with	agree
bring to a conclusion	conclude, end
come to terms with	agree, accept
due to the fact that	because
express an opinion that	affirm
for the period of	for
in such a manner that	so that
in the area / case / field of	in
in the neighborhood of	approximately
look something like	resemble
show a tendency to	tend
with reference to	regarding, about
with the result that	so

Another kind of wordiness comes from using redundant expressions—saying the same thing a second time, in different words. "Fellow colleague," "component parts," "corrosive acid," and "free gift" are phrases that contain this kind of double speech; a fellow *is* a colleague, a component *is* a part, acid *is* corrosive, and a gift *is* free. The suggested changes on the right are preferable to the redundant phrases on the left.

Redundant	Concise
absolutely essential	essential
advance reservations	reservations
basic necessities	necessities; needs
close proximity	proximity; nearness
end result	result
final conclusions/final outcome	conclusions/outcome
first and foremost	first
full and complete	full; complete
personal opinion	opinion
over and done with	over
tried and true	tried; proven

Watch for repetitious words, phrases, or clauses within a sentence. Sometimes one sentence or one part of a sentence needlessly duplicates another.

Redundant: To provide more room for employees' cars, the security department is studying ways to expand the employees' parking lot.

Edited: The security department is studying ways to expand the employees' parking lot. (Since the first phrase says nothing that the reader does not know from the independent clause, cut it.)

Adding a prepositional phrase can sometimes contribute to redundancy. The italicized words in the next list are redundant because of the unnecessary qualification they impose on the word they modify. Be on the lookout for the italicized phrases and delete them.

audible *to the ear*	hard *to the touch*
bitter *in taste*	honest *in character*
fly *through the air*	light *in weight*
orange *in color*	soft *in texture*
rectangular *in shape*	tall *in height*
second *in sequence*	twenty *in number*
short *in duration*	visible *to the eye*

Figure 2.6 shows an e-mail that Trudy Wallace wants to send to her boss about issuing smartphones to the entire sales force. Her unedited work is bloated with unnecessary words, expendable phrases, and repetitious ideas.

After careful editing, Wallace streamlined her e-mail to Lee Chadwick. Note how, in Figure 2.7 (p. 43), she pruned wordy expressions and combined sentences to cut out duplication. The revised version is only 85 words, as opposed to the 254 words in the draft. Not only has Wallace shortened her message, she has made it easier to read.

Figure 2.6   Wordy, unedited e-mail.

Wordy and unfocused subject line

| Reply | Save | Forward | Print | Delete |

To: \<lchadwick@transtech.org\>
From: \<twallace@transtech.org\>
Date: 11/7/2008 2:45 PM
Subject: Smartphones and Today's Technology

One long, unbroken paragraph is hard to follow

Repeats same idea in two or three sentences

Uses awkward and wordy sentences

Does not say what writer will do about problem

Due to the inescapable reliance on technology, specifically on e-mail and Internet communication, within our company, I believe it would be beneficial to look into the possibility of issuing smartphones, such as BlackBerry, Apple iPhone, Nokia E62, or Palm Treo, to our employees. Issuing smartphones would have a variety of positive implications for efficiency of our company. Unlike normal cell phones, these smartphones have many new features that will help our employees in their daily work, since they combine cellular phone technology with e-mail and document transfer capabilities, as well as many other important features. The employees could increase their efficiency due to the fact that they could constantly keep track of appointments on their schedule for each day. The employees would also benefit from smartphones by having access to their contacts even when they are out on the road traveling, whether they are at local office meetings or on cross-country business trips. By means of smartphones I feel quite certain that our company's correspondence would be dealt with much more speedily, since these devices will allow our employees to access their e-mail at all times. I think it would be absolutely essential for the satisfaction of our customers and to the ongoing operation of our company's business today to respond fully and completely to the possibility such a proposal affords us. It would therefore appear safe to conclude that with reference to the issue of smartphones that every means at our disposal would be brought to bear on issuing such smartphones to our employees.

The e-mail in Figure 2.6 edited for conciseness.   **Figure 2.7**

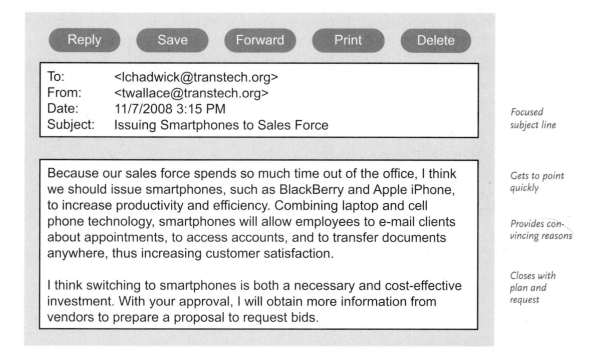

*Focused
subject line*

*Gets to point
quickly*

*Provides con-
vincing reasons*

*Closes with
plan and
request*

Guidelines for Eliminating Sexist Language

Editing involves far more than just making sure your sentences are readable. It also reflects your professional style—how you see and characterize the world of work and the individuals in it, not to mention how you want your readers to see you. Your words should reflect a high degree of ethics and honesty, free from bias and offense.

Sexist language, for example, offers a distorted, biased view of our society and discriminates in favor of one sex at the expense of another, usually women. Using sexist language offends and demeans female readers by depriving them of their equal rights. You may cost your company business if you use sexist language. Escape displaying gender bias by using inclusive language for women and men alike.

Such language is based on sexist stereotypes that depict men as superior to women. For example, calling politicians *city fathers* or *favorite sons* follows the stereotypical picture of seeing politicians as male. Such phrases discriminate against women who do or could hold public office at all levels of government.

Sexist language also prejudiciously labels some professions as masculine and others as feminine. For example, sexist phrases assume engineers, physicians, and pilots are male (*he, his,* and *him* are often linked with these professions in descriptions) while social workers, nurses, and secretaries are female *(she, her),* although members of both sexes work in all those professions. Sexist language also wrongly points out gender identities when such roles do not seem to follow biased expectations—*lady engineer, female executive, male nurse, female surgeon,* or *female astronaut.* Such offensive distinctions reflect prejudiced attitudes that you should eliminate from your writing.

Always prune the following sexist phrases: *every man for himself, gal Friday, little woman, lady of the house, the best man for the job, to man a desk or a post, the middleman, the weaker sex, woman's work, working wives,* and *young man on the way up.* Sexist terms will not only offend but also exclude many of the members of the audience you want to reach.

Ways to Avoid Sexist Language

1. Replace sexist words with neutral ones. Neutral words do *not* refer to a specific sex; they are genderless. The sexist words on the left in the following list can be replaced by the neutral nonsexist substitutes on the right.

Sexist	Neutral
alderman; assemblyman	representative
businessman	businessperson
chairman	chair; chairperson
congressman	representative
divorcée	divorced person
fireman	firefighter
foreman	supervisor
housewife	homemaker
janitress	cleaning person
landlord, landlady	owner
maiden name	family name
mailman/postman	mail carrier
man-hours	work-hours
manmade	synthetic; artificial
manpower	strength; power
policeman	police officer
repairman	repair person
salesman	salesperson
spokesman	spokesperson
woman's intuition	intuition
workman	worker

2. Watch masculine pronouns. Avoid using masculine pronouns *(he, his, him)* when referring to a group that includes both men and women.

Every worker must submit his travel expenses by Monday.

Workers may include women as well as men, and to assume that all workers are men is misleading and unfair to women. You can edit such sexist language in several ways.

a. Make the subject of your sentence plural and thus neutral.

> Workers must submit their travel expenses by Monday.

b. Replace the pronoun *his* with *the* or *a* or drop it altogether.

> Every employee is to submit a travel expense report by Monday.
> Every worker must submit travel expenses by Monday.

c. Use *his or her* instead of *his.*

> Every worker must submit his or her travel expenses by Monday.

d. Reword the sentence using the passive voice.

> All travel expenses are to be submitted by Monday.

3. Eliminate sexist salutations. Never use the following salutations when you are unsure of who your readers are:

- Dear Sir
- Gentlemen
- Dear Madam

Any woman in the audience will surely be offended by the first two and may also be unhappy with the pompous and obsolete *madam.* It is usually best to write to a specific individual, but if you cannot do that, direct your letter to a particular department or office: *Dear Warranty Department* or *Dear Selection Committee.*

Be careful, too, about using the titles *Miss, Mr.,* and *Mrs.;* sexist distinctions are unjust and insulting. It would be preferable to write *Dear Ms. McCarty* rather than *Dear Miss or Mrs. McCarty.* A woman's marital status should not be an issue. If you are in doubt, write *Dear Indira Kumar.* (Review Chapter 4, page 87, about acceptable salutations in your letters.)

4. Avoid words ending in a diminutive—*ette* or *es*—that refer to women, e.g., poetess (poet), stewardess (flight attendant), usherette (usher).

5. Never single out a person's physical appearance. Sexist physical references negatively draw attention to a woman's gender. No one would describe a male manager this way.

> The manager is a tall blonde who received her training at Mason Technical Institute.

Avoiding Other Types of Stereotypical Language

In addition to sexist language, avoid any references that stereotype an individual because of race, national origin, age, or disability. Such references almost always are irrelevant in the workplace (except for Equal Employment Opportunity Commission reports or health care studies), and are discriminatory, culturally insensitive, and unethical. To eliminate biased language in your workplace writing, follow the guidelines below.

1. Do not single out or stereotype an individual because of race, national origin, or sexual orientation.

> **Wrong:** Bill, who is African American, is one of the company's top sales reps.
> **Right:** Bill is one of the company's top sales reps.

2. Avoid words or phrases that discriminate against an individual because of age. For example, do not use *elderly, up in years, geezer, old timer, over the hill, senior moment,* or the adjectives *spry* or *frail* when they are applied to someone's age: "a spry 79." Making someone's age an issue is unfair, whatever it may be.

> **Wrong:** Jerry Fox, who will be 61 next month, comes up with obsolete plans from time to time.
> **Right:** Some of Jerry Fox's plans have not been adopted.

3. Respect individuals who may have a disability by avoiding derogatory words such as *crippled, handicapped, impaired,* or *lame* (physical disabilities) or *retarded* or *slow* (mental disabilities). Emphasize the individual instead of the physical or mental condition as if it solely determined the person's abilities.

> **Wrong:** Tom suffers from MS.
> **Right:** Tom is a person living with MS.

Keep in mind that the Americans with Disabilities Act (1990) prohibits employers from asking if a job applicant has a disability.

Also avoid using expressions that insultingly refer to a physical condition, e.g., *crippled economy, lame excuse, mental midget, insane proposal.*

Collaboration Is Crucial to the Writing Process

In the world of work, individual writing skills, such as researching, planning, drafting, revising, and editing just discussed in this chapter, are vital for your success. But you will also be expected to communicate as part of a team (including managers and co-workers) to get a report, a proposal, or even a letter written successfully. Figure 2.8 illustrates a typical workplace scenario, in which co-workers and managers come together to produce collaboratively written documents.

Collaborative writing teams benefit both employers and employees. Specific advantages of collaboration include the following.

1. **It builds on collective talents.** Many heads are better than one. A writing team profits from the diverse backgrounds and skills of its individual members. Collaboration joins individuals from diverse disciplines in the corporate world—scientists, lawyers, designers, security experts, for example.

2. **It provides productive feedback by airing diverse viewpoints.** A company wins by pooling the diverse viewpoints, constructive criticism, and immediate feedback from its work force.

3. **Team members can helpfully critique each other's suggestions, drafts, and revisions.** Playing devil's advocate, a member can help a team uncover inconsistencies or problems and correct them.

A collaborative writing team at work.   **Figure 2.8**

© Royalty-Free/CORBIS

4. Teamwork increases productivity and saves time. It cuts down on the number of meetings and conferences, saving a company time and employees travel. See how videoconferencing assists a company in doing this in Figure 10.2 (pp. 316–318).

5. Teamwork accelerates decision-making time. A group investigating problems can considerably reduce the time it takes to communicate various viewpoints and reach a decision about them.

6. Teamwork can offer psychological benefits by building confidence and morale. Working as part of a team relieves an individual of some job stress because he or she is not solely responsible for planning, drafting, and revising a document. A team, therefore, provides a safety net, assuring members that they can always talk over problems and will have help meeting deadlines.

7. Collaboration affords a greater opportunity to understand global perspectives. By working with a multinational work force, such as that described in the long report in Chapter 9 (pp. 291–308), individuals can develop greater sensitivity to and appreciation of the needs of an international audience.

Seven Guidelines for Successful Group Writing

To be successful, a collaborative writing team should observe the following seven guidelines as well as the procedures for collaboration in the writing process.

1. Understand and agree on the important goals, organization, and deadlines for the report. Everyone needs to be on the "same page" from start to finish.

2. Establish group rules early on and stick to them. Decide when and where the group will meet, how and when the group members are to communicate with each other (face-to-face, telephone, e-mail, other online technologies), how information is to be shared and with whom and how often, when various tasks are to be completed, and what "fail-safe" mechanisms are in place if and when problems arise.

3. Put the good of the group ahead of individual egos. Group harmony and productivity are essential if the report or proposal is to get done on time. Individuals need to be active participants and keen listeners. Getting one's own way can slow down the overall success of the document. In the business world, sometimes a report bears only the boss's name, not the name of the group that prepared it.

4. Agree on the group's organization. The group can appoint a leader who keeps the team on-task by being a cheerleader, a scheduler, a peacemaker who can resolve conflicts quickly, and also a referee who knows when to call time-out.[1] The leader must be skilled in interpersonal interactions. In many organizations, it is the leader who communicates with management and submits the final document.

5. Identify each member's responsibilities precisely. There must be a fair distribution of labor so that each member can use his or her particular and proven skills. One member of the group may be responsible for document design and visuals, another for research and documentation, another for writing drafts, and still another for making an oral presentation. The entire group, however, should share responsibility for the overall preparation, design, writing, and proofing of the report.

6. Provide clear and positive feedback at each meeting and for each part of the report the group prepares. Members need to come to meetings prepared, raise important questions, and make thoughtful recommendations. Not showing up, arguing, or failing to turn assignments in on time jeopardizes the entire project.

7. Follow an agreed-on timetable, but leave room for flexibility. The group should estimate a realistic time necessary to complete the various stages of their work—when drafts are due or when editing must be concluded, for example. A project schedule based on that estimate should then be prepared. **But remember: Projects always take longer than initially planned.** Prepare for a possible delay at any one stage. The group may have to submit progress reports (see pp. 246–250) to its members as well as to management.

---

[1]Adapted from Hendrie Wesigner, *Emotional Intelligence at Work* (New York: Jossey-Bass, 1997).

Sources of Conflict in Group Dynamics and How to Solve Them

The success of collaborative writing depends on how well the team interacts. Discussion and criticism are essential to discover ideas, results, and solutions. "Conflict" in the sense of conflicting opinions—a healthy give-and-take—can be positive if it alerts the group to problems (inconsistencies, redundancies, incompleteness) and provides ways to resolve them. A conflict can even help the group generate and refine ideas, thus leading to a better organized and written document.

But when conflict translates into ego tripping and personal attacks, nothing productive emerges. Everyone in the group must agree beforehand on three iron-clad working policies of group dynamics: (1) Individuals must seek and adhere to group consensus; (2) compromise may be advisable, even necessary, to meet a deadline; and (3) if the group decides to accept compromise, the group leader's final decision on resolving conflicts must be accepted.

Following are some common problems in group dynamics, with suggestions on how to avoid or solve them. (See the box on p. 50 for tips on negotiating.)

1. Resisting constructive criticism. No one likes to be criticized, yet criticism can be vital to the group effort. Be open to suggestions. Individuals who insist on "their way or no way" can become hostile to any change or revision, no matter how small.

Solution: When emotions become too heated, the group leader may wisely move the discussion to another section of the document or to another issue and allow some cooling-off time. Negotiation is an essential skill.

2. Failing to give constructive criticism. Do not saturate a meeting with nothing but negatives. You will block communication if you start criticizing with words such as "Why don't you try . . . ," "What you need is . . . ," "Don't you realize that . . . ," or "If you don't. . . ."

Solutions: When you criticize an idea, diplomatically remind the individual of the team's goals and point to ways in which revision (criticism) furthers them. Explain the problem, and offer a helpful revision. Never attack a group member. Mutual respect is everyone's right and obligation. Be objective, constructive, and cooperative.

3. Refusing to participate. Withholding your opinions hurts the group efforts; identify what you believe are major problems and give the group a chance to consider them.

Solutions: If you don't feel sure of yourself or your points, talk to another member of the group before a meeting to "test" your ideas or to write down your suggestions before a meeting to share them with the group.

4. Interrupting with incessant questions. Some people interrupt a meeting so many times with questions that all group work stops. The individual may simply be unprepared or may be trying to exercise his or her control of the group.

Solutions: When that happens, a group leader can remark, "We appreciate your interest, but would you try an experiment, please, and attempt to answer your own

---

Good and Poor Negotiators

A *good* negotiator

- Is assertive
- Is diplomatic
- Keeps details of negotiations confidential
- Remembers that everything has a price and clearly defines the exchange rate for any concession to be offered
- Works for incentives that appeal to the opponent
- Does not "dumb down" his or her behavior for the benefit of the opponent
- Abandons persuasion when talks reach the negotiation point
- Enables the opponent to save face

A *poor* negotiator

- Has a negative attitude
- Is argumentative
- Dismisses his or her own statements with self-deprecating laughs
- Criticizes freely
- Is rude
- Whines
- Attempts to buy off the opposition by conceding a point without defining what is expected in return
- Expresses statements of fact as if they were questions by raising the voice at the end of sentences
- Discusses personal issues
- Presses for an admission of error
- Fails to allow the opponent to save face

Reprinted from *Occupational Hazards* magazine (March 1999). Copyright © 1999 by Penton Media, Inc. Used by permission.

---

questions?"or if the person claims not to know, the leader might then say, "Why don't you think about it for a while and then get back to us?"

5. Inflating small details out of proportion. Some individuals waste valuable discussion and revision time by dwelling on relatively insignificant points—the choice of a single word, an optional comma—and overlook larger problems in content and organization. Nitpickers can derail the group.

Solution: If there is consensus about a matter, leave it alone and turn to more important issues. Bring the group back to the big picture.

6. Dominating a meeting. Developing interpersonal skills means sharing and responding, not taking over and being aggressive and territorial.

Solutions: The leader may say, "We need to hear from the rest of the group" or take the person aside to remind him or her of the others' right to speak. Some groups operate democratically with the "one-minute rule." Each member has one minute to make comments and does not get the floor again until everyone else has had a chance to speak.

7. Being too deferential to avoid conflict. This problem is the opposite of that described in #1. You will not help your group by being a "yes person" simply to appease a strong-willed member of the group.

Solution: Feel free to express your opinions politely; if tempers begin to flare, call in the group leader or seek the opinions of others on the team.

8. Not finishing on time or submitting an incomplete document. Meeting established deadlines is the group's most important obligation to one another and to the company. When some members are not involved in the planning stages or when they skip meetings, deadlines are invariably missed. If you miss a meeting, get briefed by an individual who was there.

Solution: The group leader can institute networking through e-mail to announce meetings, keep members updated, or provide for ongoing communication and questions. (See pp. 52–55.)

▌Collaborating Online

In today's increasingly virtual workplace, in which employees work from different offices, from home, on the road, or simply on different floors within the same building, collaborative writing often occurs online rather than face-to-face. You can also expect to collaborate on a document with individuals not employed by your company or agency, such as consultants and graphics specialists. Collaborating online takes writers into an extensive communication environment. Information flows among collaborating team members via software (also known as *groupware*) such as e-mail, instant messaging, web-based collaboration systems (Wikis), and tracking systems (such as Microsoft's "Track changes" feature). Using groupware, a collaborative team can track projects, edit, revise, and share documents, contact experts, search pertinent company files, and even create its own website.

Such software, of course, does not completely eliminate the need for a group to meet in person to discuss thorny issues, clarify subtleties, or build group harmony. But in today's technology-based, global workplace, you will certainly be expected to know how to use groupware to become a member of an online writing team.

Avoiding Problems with Online Collaboration

Regardless of the online collaborative method your team uses, it must establish ground rules by which documents are created, shared, posted, protected, and submitted. By following the guidelines listed here, your team can avoid common problems in its online collaboration:

- Be sure that every authorized team member has access to and can share all versions of the document. Team members cannot give input if they cannot open, edit, save, or share a document. To do this, every team member must use the same software.

- Verify that you and your team members are all working on the same (and correct) version of the document at the same time. Problems result when a team member wastes time and delays a deadline by editing an earlier, discarded, or otherwise incorrect version.
- Save the most recent versions of the document as well as earlier ones in separate files in case the team needs to return to these earlier copies to verify that changes have been made.
- Link each revision with the individual who made them.
- Track each contributor's changes by having an identifying color code for each one's suggested revisions or by using initials to identify who has made any changes.
- If applicable to the groupware being used, protect the document from unauthorized users or unauthorized changes by issuing team members a confidential password that might change daily or weekly.
- Require every team member to authorize or sign off on the final versions submitted to the boss, agency, or customer.

Collaborating via E-Mail

Despite the advances in technology that have streamlined and protected online collaboration (for instance, using e-rooms and tracking software), e-mail is still the most popular way collaborative writing is done on the job.

The series of e-mail exchanges in Figure 2.9 were sent among a team collaboratively writing a single document, a report on expanding a hospital parking facility. In these exchanges you will see the dynamics of collaboration through e-mail, including how a document is revised and how ethical issues are raised and resolved.

Figure 2.9   Four online collaborative writers write and revise a report via e-mail.

To:	<Ramon_Calderez@citymed.org>, <Alex_Latriere@citymed.org>, <Loretta_Bartel@citymed.org>
From:	<Nicole_Goings@citymed.org>
Date:	Wed. 7 April 2007 16:55
Subject:	Report on Expanded Hospital Parking

Now that we have a pretty clear outline and a first draft of the report before us, I think we all need to try to flesh it out. Thanks for reviewing the attached document to see how it hangs together. My initial reaction is that the draft needs reorganization and more detail.

Continued

(Continued)   **Figure 2.9**

To: <Nicole_Goings@citymed.org>, <Alex_Latriere@citymed.org>,
        <Loretta_Bartel@citymed.org>
From: <Ramon_Calderez@citymed.org>
Date: Wed. 7 April 2007 18:01
Subject: Report on Expanded Hospital Parking

Thanks for the draft. Our opening is not very convincing. We don't emphasize the hospital's reason for investing $3.2 million in expanded parking facilities. I have rewritten the opening, as you will see, and tried to link the currently inadequate parking facilities (a detriment) to the overall growth of patient care (our strong point). Then I tried to emphasize how responsive Bloomington Memorial has been to visitor needs.

Let me know what you all think.

To: <Nicole_Goings@citymed.org>, <Alex_Latriere@citymed.org>,
        <Ramon_Calderez@citymed.org>
From: <Loretta_Bartel@citymed.org>
Date: Thurs. 8 April 2007 8:23
Subject: Report on Expanded Hospital Parking

I agree with Ramon and think the revised introduction will work much better, but aren't we being too dramatic and not very pro–Bloomington Memorial by using the last sentence of his in the second paragraph—"new parking facilities will prevent visitors from walking long, bone-soaking distances in the rain." I cut it and drafted a different closing sentence.

To: <Nicole_Goings@citymed.org>, <Ramon_Calderez@citymed.org>,
        <Loretta_Bartel@citymed.org>
From: <Alex_Latriere@citymed.org>
Date: Thurs. 8 April 2007 8:54
Subject: Report on Expanded Hospital Parking

Loretta's change is OK, but is it fair to say we are adding 500 new parking spaces? The exact number is 417. I am more comfortable with saying "more than 400" or just giving the exact number. Also, I think we need another paragraph under the section now labeled "Increased Traffic Flow" and devote it entirely to Wentworth Avenue becoming a one-way street.

I inserted two points at the end of that section to make a more coherent paragraph.

To: <Alex_Latriere@citymed.org>, <Ramon_Calderez@citymed.org>,
        <Loretta_Bartel@citymed.org>
From: <Nicole_Goings@citymed.org>
Date: Thurs. 8 April 2007 11:08
Subject: Report on Expanded Hospital Parking

Alex, I am happy that you have drafted a new paragraph on Wentworth and have also done some slight editing. What do the rest of you think?

Continued

Figure 2.9   (Continued)

To: <Nicole_Goings@citymed.org>, <Alex_Latriere@citymed.org>,
           <Loretta_Bartel@citymed.org>
From: <Ramon_Calderez@citymed.org>
Date: Thurs. 8 April 2007 14:23
Subject: Report on Expanded Hospital Parking

Yes, a good job. But the more I looked at the section on "Entry Points" the more I was
troubled by including the topic of special needs under this heading. Given the fact that the
new parking facility will also involve widening the entrance to the ER, thus making 11
additional special needs spots available, is, in my mind, worthy of a separate section.
Accordingly, I think we should take special needs out of the "Entry Points" section and
create a new, even if small, section on "Special Needs."

To: <Ramon_Calderez@citymed.org>, <Nicole_Goings@citymed.org>,
           <Alex_Latriere@citymed.org>
From: <Loretta_Bartel@citymed.org>
Date: Thurs. 8 April 2007 17:17
Subject: Report on Expanded Hospital Parking

On another point, we should emphasize the new technology behind this facility, but since
the report is going to the board of directors and to other general readers, we need to cut
back on the technical descriptions, e.g., the details on stress points, pre-cast concrete,
and the low slope vehicular access ramps. I have tried to take some details out. Have I
taken out too much? Not enough? Let me hear from you!

To: <Loretta_Bartel@citymed.org>, <Ramon_Calderez@citymed.org>,
           <Alex_Latriere@citymed.org>
From: <Nicole_Goings@citymed.org>
Date: Fri. 9 April 2007 9:49
Subject: Report on Expanded Hospital Parking

Thanks and more thanks, Loretta. You are on target and so I have let your revisions stand.
BUT . . . I do think we need to retain some information about the access ramps and
covered areas surrounding them.

To: <Loretta_Bartel@citymed.org>, <Nicole_Goings@citymed.org>,
           <Alex_Latriere@citymed.org>
From: <Ramon_Calderez@citymed.org>
Date: Mon. 12 April 2007 10:23
Subject: Report on Expanded Hospital Parking

Our report must have a visual about the hospital's efforts. I e-mailed the archives earlier
this morning and they found the photograph of an aerial view of the hospital's original
parking lot in 1971. Let's incorporate that in the introduction and then use the artist's
drawing of what the new parking lot will look like to begin the section "Expanded Parking
Facilities Planned." I have scanned (and now attach) both documents.

Continued

(Continued)   **Figure 2.9**

To:	<Ramon_Calderez@citymed.org>, <Alex_Latriere@citymed.org>, <Loretta_Bartel@citymed.org>
From:	<Nicole_Goings@citymed.org>
Date:	Mon. 12 April 2007 14:04
Subject:	Report on Expanded Hospital Parking

Yes, the visuals definitely work.

I think we have done a careful job in revising and editing the report. Let me know before 4:30 today if you have any further suggestions or revisions.

Thanks for all your help. I'm going to acknowledge each of you in my cover letter to Director Fernandez.

As the example shows, many documents in the world of work are the result of collaborative efforts.

✓ Revision Checklist

- ☐ Investigated the research, drafting, revising, and editing benefits available through the computer.
- ☐ Researched my topic carefully to obtain enough information to answer all my readers' questions. Used appropriate means such as library research, online data searches, interviews, questionnaires, personal observations, site visits, or a combination of methods.
- ☐ Before writing, determined how much and what kind of information was needed to complete writing task.
- ☐ Spent sufficient time planning—brainstorming, outlining, clustering, or a combination of those techniques. Produced enough substantial material from which to shape a draft.
- ☐ Prepared enough drafts to decide on major points in message to readers. Made major changes and deletions if necessary in drafts.
- ☐ Revised drafts carefully to successfully answer readers' questions about content, organization, and tone.
- ☐ Made time to edit work so that style is clear and concise and sentences are readable and varied. Checked words to make sure they are spelled and used correctly and are appropriate for audience.
- ☐ Eliminated sexist and other stereotypical language.
- ☐ Recognized the benefits of collaborative writing for both employers and employees.

☐ Followed the guidelines for successful group writing when working on a collaborative document.

☐ Identified the sources of conflict in group dynamics and learned how to resolve them as part of a collaborative writing team.

☐ Learned how to use groupware to participate in and avoid technical problems with online collaboration.

Exercises

Additional Activities related to the writing process and collaborative writing are located at **college.cengage .com/pic/ kolinconcise2e.**

1. Below is a writer's initial brainstormed list on stress in the workplace. Revise the brainstormed list, eliminating repetition and combining related items.

> leads to absenteeism
> high costs for compensation for stress-related illnesses
> proper nutrition
> numerous stress reduction techniques
> good idea to conduct interviews to find out levels, causes, and extent of stress in the workplace
> poor morale caused by stress
> higher insurance claims for employees' physical ailments
> myth to see stress leading to greater productivity
> various tapes used to teach relaxation
> environmental factors—too hot? too cold?
> teamwork intensifies stress
> counseling
> work overload
> setting priorities
> wellness campaign
> savings per employee add up to $4,800 per year
> skills to relax
> learning to get along with co-workers
> need for privacy
> interpersonal communication
> employee's need for clear policies on transfers, promotion
> stress management workshops very successful in California
> physical activity to relieve stress
> affects management
> breathing exercises

2. Prepare a suitable outline from your revised list in Exercise 1 for a report to a decision maker on the problems of stress in the workplace and the necessity of creating a stress management program.

3. Assume you have been asked to write a short report, as Marcus Weekley did (see Figure 2.2), to a decision maker (the manager of a business you work for or have worked for; the director of your campus union, library, or security force; a city official) about one of the following topics.
   a. recruitment of international work force
   b. Internet resources
   c. security lighting
   d. food service
   e. insurance plans
   f. public transportation
   g. sporting events/activities
   h. team building
   i. morale
   j. hiring more part-time student workers

   Do some research and planning about one of those topics and the audience for whom it is intended by answering the following questions.

   - What is my precise purpose in writing to my audience?
   - What do I know about the topic?
   - What information will my audience expect me to know?
   - Where can I obtain relevant information about my topic to meet my audience's needs?

4. Using one or more of the planning strategies discussed in this chapter (clustering, brainstorming, outlining), generate a group of ideas for the topic you chose in Exercise 3. Work on your planning activities for about 15 to 20 minutes or until you have about 10 to 15 items. At this stage do not worry about how appropriate the ideas are or even if some of them overlap. Just get some thoughts down on paper.

5. The following paragraphs are wordy and full of awkward, hard-to-read sentences. Edit these paragraphs to make them more readable by using clear and concise words and sentences.
   a. It has been verified conclusively by this writer that our institution must of necessity install more bicycle holding racks for the convenience of students, faculty, and staff. These parking modules should be fastened securely to walls outside strategic locations on the campus. They could be positioned there by work crews or even by the security forces who vigilantly patrol the campus grounds. There are many students in particular who would value the installation of these racks. Their bicycles could be stationed there by them, and they would know that safety measures have been taken to ensure that none of their bicycles would be apprehended or confiscated illegally. Besides the precaution factor, these racks would afford users maximized convenience in utilizing their means of transportation when they have academic business to conduct, whether at the learning resource center or in the instructional facilities.

b. On the basis of preliminary investigations, it would seem reasonable to hypothesize that among the situational factors predisposing the Smith family toward showing pronounced psychological identification with the San Francisco Giants is the fact that the Smiths make their domicile in the San Francisco area. In the absence of contrariwise considerations, the Smiths' attitudinal preferences would in this respect interface with earlier behavioral studies. These studies, within acceptable parameters, correlate the fan's domicile with athletic allegiance. Yet it would be counterproductive to establish domicility as the sole determining factor for the Smiths' preference. Certain sociometric studies of the Smiths disclose a factor of atypicality which enters into an analysis of their determinations. One of these factors is that a younger Smith sibling is a participant in the athletic organization in question.

6. Following are very early drafts of memos that businesspeople have sent to their bosses or fellow workers. Revise and edit each draft, referring to the revising and editing checklists. Turn in your revision and the final, reader-ready copy. As you revise, keep in mind that you may have to delete and add information, rearrange the order of information, and make the tone suitable for the reader. As you edit, make sure your sentences are clear and concise and your words well chosen.

a. TO: Betty Cannales-Worth, Director
FROM: Tom Cranford
DATE: April 21, 2008
RE: Vacation request for vacation from June 12–23

I have been a highly productive employee and so I do not think that it is out of line for me to make this request. I have put in overtime and even done others' work in the department while they were away. So, I think that it is fair and just, and I can see no reason why I should not be allowed to take my vacation during the last two weeks of June.

Let me explain some of the reasons. I could have others watch my desk and do the work. I have helped them out, too, and they know it. I have been remarkably dependable. I have been readily accessible whenever their vacations have come around, and so I know it can be done.

I fully realize that this is the busiest time of the year for our company and that vacations are not usually granted during this season. But I do have personal reasons which I think should be honored/respected. Peak business times are major. I understand this, and I do hope an exception will be made in my case. After all, I do have the on-the-job training that other companies would reward. Thanks very much.

b. DATE: February 20, 2009
   TO: All Employees
   FROM: B.A. Holmes
   RE: Travel

Every company has its policies regarding travel and vouchers. Ours strike me as important and fairly straightforward. Yet for the life of me I cannot fathom why they are being ignored. It is in everyone's best interest. When you travel, you are on company time, company business. Respect that, won't you. Explain your purpose, keep your receipts, document your visits, keep track of meals.

If you see more than one client per day, it should not be too hard or too much to ask you to keep a log of each, separate, individual visit. After all, our business does depend on these people, and we will never know your true contributions on company trips unless you inform us (please!) of whom you see, where, why, and how much it costs you. That way we can keep our books straight and know that everything is going according to company policy.

Please review the appropriate pages (I think they are pages 23–25) about travel procedures. Thanks. If you have questions, give me a call, but check your procedures book or with your office/section manager, first. That will save everyone more time. Good luck.

7. The following sentences contain sexist and other biased language. Edit them to remove these errors.
   a. Every intern had to record his readings daily for the spokesman.
   b. Although Marcel's right hand was crippled, he still could use it to hunt and peck at the keyboard.
   c. She saw a woman doctor, who told her to take an aspirin every day.
   d. Our agency was founded to help mankind.
   e. John, who is a diabetic, has an excellent attendance record.
   f. Every social worker found her schedule taxing—not enough days in the week to help out man to man.
   g. Maria, who is Cuban, of course always adds spice to company events.
   h. To be a policeman, each applicant had to pass a rigorous physical and prove himself in the manly art of self-defense.
   i. Sandy Frain, the young Irish woman, called this morning wanting an appointment about the new policies on travel reimbursement.
   j. As long as his medication is adjusted just right, George Smith performs as well as the next man in this company.
   k. Sanj, a longtime member of the iTech department, is naturally adept at facts and figures.

8. A new manager will be coming to your office park in the next month, and you and five other employees have been asked to serve on a committee that will submit a report about safety problems at your office park and what should be

done to solve them. You and your team must establish priorities and propose guidelines that you want the new manager to put into practice. After two very heated meetings, you realize that what you and two other employees have considered solutions, the other half of your committee regards as the problems. Here is a rundown of the leading conflicts dividing your committee:

- **Speed bumps.** Half the committee likes the way they slow traffic down in the office park, but the other half says they are a menace because they jar car CD players.
- **Sound pollution.** Half your team wants Security to enforce a noise policy preventing employees from playing loud music while driving in and out of the office park, but the other half insists that would violate employee rights.
- **Van and sport utility vehicle parking.** Half the committee demands that vans and sport utility vehicles park in specially designated places because they block the view of traffic for any vehicle parked next to them; the other members protest that people who drive these vehicles will be singled out for less desirable parking places.

Clearly your committee has reached a deadlock and will be unproductive as long as those conflicts go unresolved. Based on the above scenario, do the following:

a. Have each person on the committee e-mail the other five committee members suggesting a specific plan on how to proceed—how the group can resolve their conflicts. Prepare your e-mail message and send it to the other five committee members and to your instructor. What's your plan to get the committee moving toward writing the report to the incoming manager?

b. Assume that you have been asked to convince the other half of the committee to accept your half's views on the three areas of speed bumps, noise control, and parking. Send the three opposition committee members an e-mail persuading them to your way of thinking. Your message must assure them that you respect their point of view.

c. Assume that the committee members reach a compromise after seeing your plan put forth in (a). Collaboratively draft a three-page report to the new manager.

CHAPTER 3

Writing Memos, Faxes, and E-Mails

Memos, faxes, and e-mail messages are the types of writing you will prepare most frequently on the job. These three forms of business correspondence are quick, easy, and effective ways for a company to communicate internally as well as externally. You will find yourself preparing one or more of these types of writing each day to co-workers in your department, to colleagues in other departments and divisions of your company, to decision makers at all levels, and to clients and customers as well.

To expand your understanding of memos, faxes, and e-mails, take advantage of the Web Links, Additional Activities, and ACE Self-Tests at **college.cengage .com/pic/ kolinconcise2e.**

■ What Memos, Faxes, and E-Mails Have in Common

1. Each of these forms of writing is streamlined for the busy world of work. Memos are far less formal in tone than letters; e-mails can be even more informal than a memo. Memos, faxes, and e-mails also require you to follow different formats than you do for letters.

2. They give busy readers information fast. While these messages can be about any topic in the world of work, most often they focus on the day-to-day activities and operations at your company—sales and product information, policy and schedule changes, progress reports, orders, and so on.

3. Even though routine, they still demand a great deal of thought and time. Which one you use—memo, fax, or e-mail—depends on your company policy, the nature of your message, and your audience's needs and expectations. While some individuals believe we are moving toward a paperless office, most companies will still want to see a paper or electronic trail documenting what has been done when and by whom. Your success as an employee can depend as much on your preparing a readable and effective memo, fax, or e-mail as it will on your technical expertise.

Memos

Memorandum, from which the term *memo* comes, is a Latin word for "something to be remembered." The Latin meaning points to the memo's chief function: to record information of immediate importance and interest in the busy world of work. **Memos** are brief in-house correspondence sent up and down the corporate ladder. Employees send memos to their supervisors, and workers send memos to one another. Figure 3.1 shows a memo sent from one worker to another. Figure 3.2 illustrates a memo sent from the top down; Figure 3.3 shows a memo sent from an employee to management. Memos are not as formal as letters and contain the terms and abbreviations familiar to employees of your company. They can be on paper, or sent through e-mail as in Figure 1.7 (p. 17) or Figures 3.4 and 3.5 (pp. 70 and 71).

Figure 3.1   Standard memo format.

Header

Memo parts

Introduction

Preview

Body: Numbered list helps reader follow information quickly

Conclusion: Asks for comments

MEMO

TO: Lucy
FROM: Roger
DATE: November 12, 2008
SUBJECT: Review of Successful Website Seminar

As you know, I attended the "How to Build a Successful Website" seminar on November 7 and learned the "rules and tools" we will need to redesign our own site.

Here is a review of the major topics covered by the director, Jackie Chen:

1. Keep your website content-based—hit your target audience.
2. Visualize and "map out" your site's links.
3. Design your website to look the way you envision it—make it aesthetically pleasing.
4. Make your site easy to navigate.
5. Maintain your page outline; make changes when necessary.
6. Create hot links and image maps to move users from page to page.
7. Add appropriate animation using software such as Adobe®, Flash®, or Swish Max.

Could we meet in the next day or two to discuss recreating our own website in light of these guidelines? I would really like your suggestions about this project. Thanks.

Memo with a clear introduction, discussion, and conclusion.   **Figure 3.2**

Dearborn Equipment Company

To: Machine Shop Employees
From: Janet Hempstead, Shop Supervisor
Date: September 27, 2008
Subject: Cleaning Brake Machines

During the past two weeks I have received several reports that the brake machines are not being cleaned properly after each use. Through this memo I want to emphasize and explain the importance of keeping these machines clean for the safety of all employees.

Introduction gives purpose and importance of memo

When the brake machines are used, the cutter chops off small particles of metal from brake drums. These particles settle on the machines and create a potentially hazardous situation for anyone working on or near this equipment. If the machines are not cleaned routinely before being used again, these metal particles could easily fly into an individual's face when the brake drum is spinning.

Discussion states why problem exists and how to solve it

To prevent accidents like this from happening, please make sure you vacuum the brake machines after each use.

Safety message is boldfaced for emphasis

You will find two vacuum cleaners for this purpose in the shop—one of them is located in work area 1-A and the other, a reserve model, is in the storage area. Vacuuming brake machines is quick and easy. It should take no more than a few seconds. This is a small amount of time to make the shop safer for all of us.

Encourages participation

Thanks for your cooperation. If you have any questions, please call me at Extension 324 or come by my office.

Conclusion builds goodwill and asks for questions

204 South Mill St., South Orange, NJ 02341-3420 (609) 555-9848 JHEMP@dearco.com

Figure 3.3   A memo that uses headings to highlight organization.

RAMCO INDUSTRIES

Where Technology Shapes Tomorrow
ramco@gem.com   http://www.Ramcogem.com

TO:	Rachel Mohler, Vice President
	Harrison Fontentot, Public Relations
FROM:	Mike Gonzalez MG
SUBJECT:	Ways to Increase Ramco's Community Involvement
DATE:	March 2, 2009

At our planning session in early February, our division managers stressed the need to generate favorable publicity for our new Ramco facility in Mayfield. Knowing that such publicity will highlight Ramco's visibility in the area, I think the company's image might be enhanced in the following ways.

CREATE A SCHOLARSHIP FUND

Ramco would receive favorable publicity by creating a scholarship at Mayfield Community College for any student interested in a career in technology. A one-year scholarship would cost $4,800. The scholarship could be awarded by a committee composed of Ramco executives and staff. Such a scholarship would emphasize Ramco's support for technical education at a local college.

OFFER SITE TOURS

Guided tours of the Mayfield facility would introduce the community to Ramco's innovative technology. The tours might be organized for community and civic groups. Individuals would see the care we take in production and equipment choices and the speed with which we ship our products. Of special interest to visitors would be Ramco's use of industrial robots alongside its employees. Since these tours would be scheduled well in advance, they should not conflict with our production schedules.

PROVIDE GUEST SPEAKERS

Many of our employees would be excellent guest speakers at civic and educational meetings in the Mayfield area. Possible topics include the technological advances Ramco has made in designing and engineering and how these advances have helped consumers and the regional economy.

Thanks for giving me your comments as soon as possible. If we are going to put one or more of the suggestions into practice before the facility opens, we'll need to act before the end of the month.

Company logo

Introduction supplies background and rationale

Headings reflect organization

Body offers concrete evidence (costs, location, personnel) that plan can work

Closing requests feedback and authorization

Memo Protocol and Company Politics

Memos reflect a company's image—its politics, policies, and organization. Note how Janet Hempstead's memo in Figure 3.2 reminds workers about a crucial safety policy at Dearborn Equipment Company. A company's logo may even appear on the top, as in Figures 3.2 and 3.3. These memos reflect the company's corporate culture and interests and the ways in which they build employee morale and encourage productivity.

Regardless of where you work, your employer will expect memos to be timely (don't wait until the day of a meeting to announce it), professional, and tactful. Just because a memo is an informal in-house communication does not mean you can be gruff, curt, or bossy. Politeness and diplomacy count a lot at work. Learning effective memo writing is vital to your success in any organization.

Most companies have their own memo *protocol*—accepted ways in which in-house communications are formatted, organized, written, and routed. In fact, some companies offer seminars on how employees are to prepare communications. In the corporate world, protocol determines where your memo will go. In addition, use common sense. Don't send copies of memos to people who don't need them (it wastes paper and unnecessarily wastes the time of those who don't need to see them) or to high-ranking company personnel instead of your immediate supervisor, who may think you are going over his or her head.

Functions of Memos

Memos serve a variety of functions, including

- announcing a company policy or plan
- changing a policy or procedure
- offering information (FYI) or making a request
- explaining a procedure or giving instructions
- clarifying or summarizing an issue
- alerting readers to a problem or to a deadline
- confirming the outcome of a conversation
- calling a meeting
- providing documentation necessary for business
- offering suggestions or recommendations
- documenting, for your own protection, something you did or did not do
- summarizing a long report or proposal

Memos are valuable written records used for a variety of purposes. They are used for a number of short reports; see Chapter 8 for examples of periodic, field trip, or progress reports in memo format. Many internal proposals (pp. 261–267) also are written as memos. In fact, the memo in Figure 3.3 is an example of such a proposal.

Memo Format

Memos vary in format and the way they are sent. Some companies use standard, printed forms (Figure 3.1), while others have their names (letterhead) printed on their memos (as in Figures 3.2 and 3.3). You can also create a memo by including the necessary parts in an e-mail, as in Figures 3.4 and 3.5, which appear later in this chapter.

As you can see from looking at Figures 3.1 through 3.3, memos look different from letters, and they are less formal. Because they are often sent to individuals within your company, memos do not need the formalities necessary in business letters, such as an inside address, salutation, complimentary close, or signature line, as discussed in Chapter 4 (see pp. 83–88).

Basically, the memo consists of two parts: the identifying information at the top and the message itself. The identifying information includes these easily recognized parts: **To, From, Date,** and **Subject** lines.

```
TO:        Aileen Kelly, Chief Computer Analyst
FROM:      Stacy Kaufman, Operator, Level II
DATE:      January 30, 2009
SUBJECT:   Progress report on the fall schedule
```

You can use a memo template in your word processing program that will list these headings, as follows, to save time.

```
TO:
FROM:  Linda Cowan
DATE:  October 4, 2008
RE:    (Enter subject here.)
```

On the **To** line, type the name and job title of the individual(s) who will receive your memo or a copy of it. If your memo is going to more than one reader, make sure you list your readers in the order of their status in your company or agency, as Mike Gonzalez does in Figure 3.3 (according to company policy the vice president's name appears before that of the public relations director). If you are on a first-name basis with the reader, use just his or her first name, as in Figure 3.1. Otherwise, include the reader's first and last names.

On the **From** line, type your name (use your first name only if your reader refers to you by it) and your job title (unless it is unnecessary for your reader). Some writers handwrite their initials after their typed name to verify that the message comes from them.

On the **Subject** line, type the purpose of your memo. The subject line serves as the title of your memo; it summarizes your message. Vague subject lines, such as "New Policy," "Operating Difficulties," or "Shareware," do not identify your message precisely and may suggest that your message is not carefully restricted or developed. Note how Mike Gonzalez's subject line in Figure 3.3 is so much more precise than just saying "Ramco's Community Involvement."

On the **Date** line, do not simply name the day of the week—Friday. Give the full calendar date—June 4, 2009.

Memo Style and Tone

The audience within your company will determine your memo's style and tone (for a review of identifying audience, see Chapter 1, pp. 4–9). When writing to a co-worker whom you know well, you can adopt a casual, conversational tone. You want to be seen as friendly and cooperative. In fact, to do otherwise would make you look self-important, stuffy, or hard to work with. Consider the friendly tone appropriate for one colleague writing to another in Roger's memo to Lucy in Figure 3.1. Note how he ends in a polite but informal way.

When writing a memo to a manager, though, you want to use a more formal tone than you would when communicating with a co-worker or peer. Your boss will expect you to show a more respectful, even official, posture. See how formal yet conversationally persuasive Mike Gonzalez's memo to his bosses is in Figure 3.3. His tone and style are a reflection of his hard work as well as his respect for his employer. Here are two ways of expressing the same message, the first more suitable when writing to a co-worker and the second more appropriate for a memo to the boss.

> **Co-worker:** I think we should go ahead with Marisol's plan for reorganization. It seems like a safe option to me, and I don't think we can lose.
>
> **Boss:** I think that we should adopt the organizational plan developed by Marisol Vega. Her recommendations are carefully researched and persuasively answer all the questions our department has about solving the problem.

When a boss writes to workers informing them about policies or procedures, as Janet Hempstead does in Figure 3.2, the tone of the memo is official and straightforward. Yet even so, Hempstead takes into account her readers' feelings (she does not blame) and safety, which are at the forefront of her rhetorical purpose.

Finally, remember that your employer and co-workers deserve the same clear and concise writing and attention to the "you attitude" (see Chapter 4, pp. 00–00) that your customers do. Memos require the same care and should follow the same rules of effective writing as letters do.

Strategies for Organizing a Memo

Organize your memos so that readers can find information quickly and act on it promptly. For longer, more complex communications, such as the memos in Figures 3.2 and 3.3, your message might be divided into three parts: (1) introduction, (2) body, or discussion, and (3) conclusion. Regardless of how short or long your memo is, recall the three *P*'s for success—*p*lan what you are going to say; *p*olish what you wrote before you send it; and *p*roofread everything.

Introduction
The introduction of your memo should do the following:

- Tell readers clearly about the issue or policy that prompted you to write.
- Explain briefly any background information the reader needs to know.
- Be specific about what you are going to accomplish in your memo.

Do not hesitate to come right out and say, "This memo explains new e-mail security procedures" or "This memo summarizes the action taken in Evansville to reduce air pollution."

Body (Discussion)

In the body, or discussion section, of your memo, help readers in these ways:

- State why a problem, procedure, or decision is important, who will be affected by it, and what the consequences are.
- Indicate why and what changes are necessary.
- Give precise dates, times, locations, and costs.

See how Janet Hempstead's memo in Figure 3.2 carefully describes an existing problem and explains the proper procedure for cleaning the brake machines.

Conclusion

In your conclusion, state specifically how you want the reader to respond to your memo. To get readers to act appropriately, you can do one or more of the following:

- Ask readers to call you if they have any questions, as in Figure 3.2.
- Request a reply—in writing, over the telephone, via e-mail, or in person—by a specific date, as in Figures 3.1 and 3.3.
- Provide a list of recommendations that the readers are to accept, revise, or reject, as in Figures 3.1 and 3.3.

Organizational Markers

Throughout your memo use the following organizational markers, where appropriate:

- **Headings** organize your work and make information easy for readers to follow, as in Figure 3.3.
- **Numbered** or **bulleted lists** help readers see comparisons and contrasts readily and thereby comprehend your ideas more quickly, as in Figure 3.1.
- **Underlining** or **boldfacing** emphasize key points (see Figure 3.2). Do not overuse this technique; draw attention only to main points and those that contain summaries or draw conclusions.

Organizational markers are not limited to memos; you will find them in e-mail, letters, and reports as well. (See Chapter 6, pp. 169–178.)

Faxes

Even though you may write e-mails extensively, faxes (facsimile) are still in widespread use to send copies of original letters, memos, reports, graphs, blueprints, and artwork over phone or computer lines. You also can send graphics that resemble actual photos. Send a fax when your reader needs an exact copy of an original document, for example, a contract, letter, proposal, or map. A fax does not take the place of the original, which you may later need to send via snail, certified, or express mail.

Faxes are especially effective if you have to make a few changes in a detailed document (for example, a contract or a boilerplate) and do not want to rekey or transmit the entire work. Unlike e-mail, faxes provide a signature authorization and give recipients a hard copy.

Be sure to use a fax cover sheet, which lists the person(s) sending and receiving the fax; their addresses and phone and fax numbers; and the total number of pages being faxed.

Be aware that, unless the recipient has her or his own secured fax machine, your confidentiality is not easily protected when communicating by fax. If your fax is sent to a machine available to the entire office staff, anyone can read it.

Fax Guidelines

When you send a fax, observe the following guidelines:

1. Because the type size of a document generally is reduced during transmission, print your fax message in a larger point size (12 or 14).
2. Avoid writing any comments at the very top or bottom of a fax. Your notes might be cut off or blurred in transmission.
3. Make sure the document you are faxing is clear, but always include your phone and fax numbers in case the recipient needs to verify your message.
4. Be careful about sending anything longer than three or four pages. You might tie up the recipient's phone line. For a longer document, call before you fax to see whether the recipient will allow you to fax it.

E-Mail

E-mail is the most common workplace communication. It is the lifeblood of every business or organization because it expedites communications within your firm as well as communications with those outside of it. Professionals in the world of work may receive hundreds of e-mails a day from supervisors, colleagues, clients, and vendors. These e-mails may contain pictures; video clips; soundbites; and various tables, lists, and statistical files. As we saw, memos can be sent via e-mail.

Business E-Mail Versus Personal E-Mail

E-mail is the most informal, relaxed type of business correspondence, far more so than a printed memo, letter, short report, or proposal. (However, e-mail should never be sent in place of a formal letter.) Think of business e-mail as a polite, informative, and professional phone conversation—friendly, to the point, and always accessible. Yet, even though business e-mail is basically informal and casual, sending it does not mean you can forget about your responsibilities as an informed, cautious, and courteous writer. Review Figure 2.9, which exemplifies how e-mail can be used as a tool in collaborative writing.

The e-mail you write on the job will require more from you as a writer than your personal e-mail will. You cannot be sloppy or unorganized. Proofread and, if necessary, revise your e-mail before you send it. You need to follow all the rules of proper spelling, capitalization, punctuation, and word choice, plus the e-mail guidelines in the following section. The **tone** (see pp. 66–67) of your business e-mail should be much more professional than the instant messaging you may do with friends or conversations you may have in a chat room. You cannot write to your employer or a customer the way you would to an old friend.

Unlike your personal e-mail, your business e-mail must consider the impact it will have on your company and on your career. When you send a business e-mail, you are representing more than just yourself and your preferences, as in a personal e-mail. You are speaking on behalf of your employer. Since your e-mail must reflect your company's best image, make sure it is businesslike, carefully researched, and polite. Sarcasm, slang, and aggressive name calling do not belong in a company e-mail. Figures 3.4 and 3.5 are examples of effectively written business e-mails. Notice that e-mail can be cordial without being unprofessional.

Figure 3.4   An example of an effectively written business e-mail.

Header contains all necessary information

Gives all necessary details concisely

Indicates follow-up

Uses informal yet professional tone

| Reply | Save | Forward | Print | Delete |

To: <marge.parish@craftworks.com>
From: <Ali.Fatoul@craftworks.com>
Date: 27 July 2009 9:38
Subject: Status of Hinson-Davis Order

Hi, Marge,

At last the Hinson-Davis Company received its order, and they are very pleased with our service. Victor Arana, their district manager, just called me to say the order came in at 9:00 a.m. and by 9:15 it was up on their system.

I am going to send Arana a thank-you letter today to keep up the goodwill.

Things could not have gone smoother. Congrats to all.

Thanks,

Ali

E-mail sent to a distribution list of co-workers.   **Figure 3.5**

Reply	Save	Forward	Print	Delete

To:	<annulla.cranston@netech.com>; <peter.hwang@netech.com>; <margaret.habermas@netech.com>; <a.pena@netech.com>
From:	<melinda.bell@netech.com>
Date:	19 June 2009 13:18
Subject:	Collaboration on annual report

Hello, Team:

To follow up on our conversation yesterday regarding working on this year's annual report, I'm glad our schedules are flexible. I've checked our calendars and we are all available next Tuesday the 26th at 10:30 a.m. Let's meet in Conference Room 410.

Don't forget we have to draft a two- to three-page overview first that explains Northeast's strategic goals and objectives for fiscal year 2008. Not an easy assignment, but we can do it, gang.

It would be a big help if Annulla would bring copies of the reports for the last three years. Would Peter please call Ms. Jhandez in Engineering for a copy of the speech she gave last month to the Powell Chamber of Commerce? If memory serves me correctly, she did a first-rate job summarizing Northeast's accomplishments for 2007.

Thanks for all your splendid work, team. See you Tuesday.

Melinda Bell
Director, Marketing
New Tech
<melinda.bell@netech.com>
FAX: (603) 555-2162
Voice: (603) 555-1505
http://www.netech.com

Starts with context for and confirmation of meeting

Provides clear explanations and instructions

Requests information politely

Ends by building morale

Provides contact information

E-Mails Are Legal Records

Employers own their internal e-mail systems and thus have the right to monitor what you write and to whom. Any e-mail written at work on the company's server can be saved, stored, forwarded, and, most significantly, intercepted. You can be fired for writing an angry or abusive e-mail. Keep in mind your e-mail can easily be converted into an electronic paper trail. You never know who will receive and then forward your e-mail—to your boss, an attorney, or a licensing board. Many companies include disclaimers protecting themselves from legal action against them because of an employee's offensive behavior in a company e-mail.

Here are some good rules to follow about your company's e-mail:

- Do not use it for your personal messages. Use it only for appropriate company business, and make sure you are professional and conscientious.
- Never write an e-mail to discuss a confidential subject—a raise, a grievance, or a complaint about a co-worker. Meet with your supervisor in person.
- Make sure of your facts before sending an e-mail to customers. If it gives them wrong or misleading information about prices, warranties, or safety features, your company can be legally liable.

Guidelines for Using E-Mail

Using e-mail technology at work obligates you to prepare and organize your messages carefully with your specific reader's needs in mind as well as those of his or her company and your own organization. Following the guidelines below will help you write effective business e-mail messages.

1. **Make sure your e-mail is confidential and ethical.**

    a. Do not send or forward chain letters, political ads, cartoons, etc.
    b. Do not reply to spam, which will only multiply the amount of it you will receive.
    c. Never attack your employer, a colleague, a customer, or any firm.
    d. Avoid *flaming*, that is, using strong, angry language that mocks, attacks, or insults your reader, as in Figure 3.6 (p. 76). Abusive, obscene, or racially offensive language in an e-mail constitutes grounds for dismissal.
    e. Send nothing through e-mail that you would not want to see on your company's website, intranet, or on the front page of your newspaper.
    f. Be cautious when you choose Reply All, so that you do not send something that is embarrassing or illegal.

2. **Observe all of the proprietary requirements when using e-mail.**

    a. Use quotation marks to indicate words that are not yours, and always give a source.
    b. Do not forward a co-worker's or boss's e-mail without that person's approval.
    c. Be careful not to change the wording of a message that you are expected simply to read and then forward.

> **d.** Print e-mails that are necessary for a presentation or report, or those that contain directions you may need to refer to again.
>
> **e.** Copy yourself on e-mails where you have to establish or maintain a paper trail.

3. Use an acceptable format.

Since most e-mail is read online rather than in hard copy, consider how your message will look on a screen and the needs of your online reader.

> **a. Make your e-mail easy to read.**
>
> - Do not send e-mail in all capital letters. It's hard to read, unprofessional, and looks as if you are shouting. Conversely, don't write in all lower-case letters, which implies a lack of knowledge of punctuation.
> - Use no more than 65 characters per line and format your e-mail so that it wraps the text, preventing your e-mail from looking like a poem.
> - Break your message into paragraphs. A screen filled with one long, unbroken paragraph is intimidating.
> - Watch the length of your paragraphs. Keep them to three or four lines and double-space between them.
> - Avoid using boldface, asterisks, italics, and underlining, which may not be translatable by your recipient's e-mail system and garble or distort your message.
>
> **b. Make your e-mail easy to process.**
>
> - Include all parts of your message—correct address, subject line, date, and so on.
> - Avoid vague one-word subjects. A subject line like "Bill" would leave your reader wondering if your e-mail is about a person, an unpaid account, or a notice just sent.
> - Don't use highly emotional words like "Urgent," "Important," "Read at Once" in your subject line. They will turn your readers off.
> - Get to the point right away. Because your readers receive a lot of e-mail, they may look only at the first few lines you write.
> - Refer to any relevant previous e-mail to give busy readers necessary background and context information quickly.
> - Include a *signature block* with your name, company affiliation, title, address, and phone numbers, as in Figure 3–5.

4. Follow all of the rules of "netiquette" when answering e-mail.

> **a. Respond promptly to your e-mail.**
>
> - Check your mail several times each day—don't let it build up.
> - Reply the same day, if possible. Let correspondents know you received their messages.

- Prioritize your messages so that you reply to your boss first and to others later.
- If you expect a delay in getting information, tell the correspondent how long it will be before you can respond.
- If you are offline, use your e-mail software's **auto-reply feature** to send back a message to that effect (and indicate when you will return) to each correspondent.
- Keep your address book up-to-date.

b. Identify your audience correctly.

- Send your e-mail to the right address (is it an individual or a group?).
- Verify if your reader wants unsolicited mail.
- Learn all you can about your reader to shape the tone of your mail to his or her needs; judge how much information the reader needs.
- Do not inflate your distribution list by clicking on Reply All with each e-mail. Send messages only to readers who need them (e.g., scheduling a meeting, as in Figure 3.5). Managers dislike being included on unnecessary distribution lists because it wastes their time.

c. Be courteous to your reader.

- Don't send the same message over and over. Consider any time zone differences between you and your reader. For example, it might be 2 a.m. when your e-mail arrives, so don't expect an immediate reply.
- Respect the cultural traditions of non-native speakers of English. (See Chapter 4, pp. 117–125.) Do not use first names unless the reader approves, and stay away from jokes and slang. Avoid including abbreviations, monetary units, and measurements that your reader may not know.
- Delete long strings of previously answered messages when you reply so that the reader won't waste time scrolling to find your answer. Put your message first if you are forwarding an e-mail.
- Choose the New Message option to avoid sending old news again and again.
- Do not send long attachments or graphics files without obtaining recipients' permission. Such messages may be difficult to read onscreen or use a lot of memory. If your message is long, send it as an attachment.

5. **Adopt a professional style.**

   a. Use a salutation (greeting) or complimentary close (farewell), but always follow your company's policy:

   - to a colleague—Hi, Hello; Bye, Thanks
   - to a customer—Dear Ms. Pietz; Dear Bio Tech; Sincerely

   b. Keep your message concise.

   - Cut wordy phrases. See how many words you can eliminate without distorting your message.

- Don't turn your e-mail into a telegram. "Report immediately: need for meeting" makes you sound demanding. Likewise, responding with only a "Yes" or "No" is discourteous. Save "Yeah," "Nope," and "Huh" for your personal e-mails.
- Send only the information needed to answer the reader's questions or concerns. Exclude nonessential details and chatter.

   c. Avoid abbreviations (e.g., BAK, back at keyboard; OTOH, on the other hand), catchwords, or phrases people outside your office might not understand or appreciate.

   d. Do not use jargon unless it is appropriate for the context and your audience.

   e. Avoid using **emoticons** (e.g., happy [:)], or sad [: ( ] faces) made with punctuation marks, which are unprofessional.

   f. End your message politely. Let the reader know you appreciate and welcome his or her help.

6. **Ensure that your e-mail is safe and secure.**

   a. Use various e-mail protection services and software.

   b. Avoid obtaining e-mail viruses by deleting unopened, unsolicited e-mail attachments. Never forward an infected e-mail to anyone else.

   c. Don't be a victim of identity theft e-mail, or "phishing" where someone claims to be a legitimate company and asks for personal information, such as your bank account or social security number. Legitimate companies will **not** contact you via e-mail for such information.

   d. Do not give out your e-mail address unless you know how it's going to be used.

   e. Never provide company financial information unless you are sure the information will be relayed over a safe connection. Check with your boss before providing such information.

Figure 3.6 shows an example of a poorly written e-mail that violates many of the preceding guidelines. Figure 3.7 contains an effective revision that reflects the professional and courteous way the writer and his company do business.

E-Mail Compared with Other Business Communications

Table 3.1 compares and contrasts the function, scope, and format of e-mail with memos, letters, and faxes. Note that e-mails are brief, informal, and to-the-point messages that are the workhorses in the world of business. They should never take the place of far more formal and official documents. Always acknowledge a business gift or courtesy by sending a handwritten note of thanks rather than just dashing off an e-mail. International readers, in particular, will expect this. Keep this table in mind as you learn more about various types of letters, including business letters, job application letters, and reports covered in the following chapters.

Figure 3.6   A poorly written e-mail.

No specific
reader

Vague subject
line

Unprofessional
greeting

Discourteous
tone
All caps shout
Unprofessional
emoticon
Flaming

Insufficient
information

Unclear
abbreviation
Unprofessional
emoticon
No signature
block

| Reply | Save | Forward | Print | Delete |

To: \<newtech@widedoor.com\>
From: \<sammy@dataport.com\>
Date: 11/19/08 09:50 (CST)
Subject: Upgrades

HEY GUYS------------

ARE YOU AWAKE OUT THERE? THIS IS THE THIRD TIME I HAVE
SENT THIS MESSAGE. AND I NEED YOU TO GET BACK TO ME
STAT. MY BOSS IS ON MY BACK. : P

I NEED THE UPGRADES YOUR SALES FOLKS--ROBERT T., JAN W.,
AND GRAF H.--PROMISED BUT NEVER MADE GOOD ON.

FWIW YOU HAVE MISSED THE BOAT. ☹

SAMMY

A revised, effective version of the poor e-mail in Figure 3.6.   **Figure 3.7**

| Reply | Save | Forward | Print | Delete |

To: \<MWood@widedoor.com>
From: \<sammy@dataport.com>
Date: 11/19/08
Subject: Upgrades for service contract #4552

Specific e-mail address

Precise subject

Hello, Mary:

I would appreciate your delivering the upgrades for our service contract #4552 by Thursday afternoon, the 22nd of November, if at all possible.

We need to proceed to the next phase of our operation and the upgrades are crucial to that task.

I am attaching a copy of our service agreement with Wide Door for your convenience.

If you run into any problem with the delivery date, please give me a call this afternoon or e-mail me.

Thanks,

Sammy

Samuel Atherton
Operations Assistant
Data Port
4300 Morales Highway
San Padre, CA 95620-0326
Voice mail: 723-555-1298
http://dataport.com

Polite salutation

Gets to the point concisely

Provides explanation and documentation

Ends with clear-cut directions

Professional close

Includes signature block

TABLE 3.1   The Uses of E-Mails, Memos, Letters, and Faxes

	E-Mail	Memo	Letter	Fax
Brief messages	X	X		
Informal	X	X		X
Formal			X	X
Legal record		X	X	X
Relaxed tone	X	X		
Confidential material		X	X	
Secure site		X	X	
Multiple pages			X	X
Reports		X	X	X
In-house messages	X	X		
Proofreading	X	X	X	

Revision Checklist

Memos
- [ ] Used appropriate and consistent format.
- [ ] Announced purpose of memo early and clearly.
- [ ] Organized memo according to reader's need for information, with main ideas up front; supplied clear conclusion.
- [ ] Made style and tone of memo suitable for audience.
- [ ] Included bullets, lists, underscoring where necessary to reflect logic and organization of memo and for ease of reading.
- [ ] Refrained from overloading reader with unnecessary details.

Faxes
- [ ] Verified reader's fax number.
- [ ] Sent cover sheet with number of pages faxed and phone number to call in the event of transmission trouble.

☐ Enlarged font to minimize reduction of type in transmission.
☐ Excluded anything confidential or sensitive if reader's fax machine is not secure.
☐ Promptly returned any calls regarding transmission difficulties with fax.

E-Mails
☐ Considered the differences between business e-mail and personal e-mail.
☐ Included my essential contact information in a signature block.
☐ Ensured business e-mail was confidential and ethical.
☐ Observed proprietary requirements of workplace e-mail.
☐ Used acceptable e-mail format at work.
☐ Followed the rules of "netiquette" in all e-correspondence.
☐ Adopted a professional e-mail style at work.
☐ Made sure workplace e-mails were safe and secure.

Exercises

1. Write a memo to your boss saying that you will be out of town two days next week and three days the following week for **one** of the following reasons: (a) to inspect some land your firm is thinking of buying, (b) to investigate some claims, (c) to look at some new office space for a branch your firm is thinking of opening in a city five hundred miles away, (d) to attend a conference sponsored by a professional society, or (e) to pay calls on customers. In your memo, be specific about dates, places, times, and reasons.

Additional Activities related to memos, faxes, and e-mails are located at **college.cengage.com/pic/kolinconcise2e.**

2. Select some change (in policy, procedure, schedule, personnel assignment) you encountered on a job held in the last two or three years and write an appropriate memo describing that change. Write the memo from the perspective of your former employer explaining the change to employees.

3. Write an e-mail to a business that provides daily or weekly information to interested customers and submit its response along with your e-mail request to your instructor. Choose one of the following:
   a. an airline: an up-to-date schedule along a certain route and information about any bonus-mile or discount programs
   b. a catalog order company: information about any weekly specials for Internet users
   c. a stock brokerage firm: free quotes or research about a particular stock
   d. a resort: special rates for a given week

4. Write an e-mail with one of the following messages, observing the guidelines discussed in this chapter.
   a. You have just made a big sale and you want to inform your boss.

b. You have just lost a big sale and you have to inform your boss.

c. Tell a co-worker about a union or national sales meeting.

d. Notify a company to cancel your subscription to one of its publications because you find it to be dated and no longer useful in your profession.

e. Request help from a listserv about research for a major report you are preparing for your employer.

f. Advise your district manager to discontinue marketing one of the company's products because of poor customer acceptance.

g. Send a short article (about 200 words) to your company newsletter about some accomplishment your office, department, or section achieved during the last month.

h. Write to a friend studying finance at a German, Korean, or South American university about the biggest financial news in your town or neighborhood in the last month.

5. As a collaborative venture, join with three or four classmates to prepare one or more of the e-mail messages for Exercise 4. Send each other drafts of your messages for revision. Submit the final copy of the group's effort.

6. Rewrite the following e-mail to make it more suitable.

Hi––

This new territory is a pain. Lots of stops; no sales. Ughhhh. People out here resistant to change. Could get hit by a boulder and still no change. Giant companies ought to be up on charges. Will sub. reports asap as long as you care rec.

The long and short of it is that market is down. No news=bad news.

7. Send your instructor an e-mail message about the project you are now working on for class, outlining your progress and describing any difficulties you are having.

8. You have just missed work or a class meeting. E-mail your employer or your instructor explaining the reason and indicate how you intend to make up the work.

CHAPTER 4

Writing Letters

Letters are among the most important and official writing you will do on your job. Businesses take letter writing very seriously, and employers will expect you to prepare and respond to your correspondence effectively. Your signature on a letter tells readers you are accountable for everything in it. The higher up the corporate ladder you climb, the more letters you will be expected to write.

Because letter writing is so significant to your career, this chapter introduces you to the entire process, provides guidelines and problem-solving strategies, and shows you how to prepare the most frequent types of business letters. It also shows you how to write for international readers.

<aside>To expand your understanding of writing letters, take advantage of the Web Links, Additional Activities, and ACE Self-Tests at **college.cengage.com/pic/kolinconcise2e.**</aside>

Letters in the Age of the Internet

Even in this age of electronic communications, letters are still vital for the following reason in the world of work:

1. **Letters represent your company's public image and your competence.** A firm's corporate image is on the line when it sends a letter. Carefully written letters can create goodwill; poorly written letters can anger customers, cost your company business, and project an unfavorable image of you.

2. **Letters are far more formal—in tone and structure—than any other type of business communication.** Memos and e-mail are the least formal communications.

3. **Letters constitute an official legal record of an agreement.** They state, modify, or respond to a business commitment. When sent to a customer, a signed letter constitutes a legally binding contract. Be absolutely sure that what you put in a letter about prices, warranties, equipment, delivery dates, and/or other issues is accurate. Your reader can hold you and your company to such written commitments.

4. **Unlike e-mail, many businesses require letters to be routed through channels before they are sent out.** Because they convey how a company looks and what it offers to customers, letters often must be approved at a variety of corporate levels.

5. Letters are more permanent than e-mails. They provide a documented hard copy. Unlike e-mails that can be erased, letters are often logged in, filed, and bear a written authorized signature. They are also far more confidential.

6. A letter is still the official and expected medium through which important documents and attachments (contracts, specifications, proposals) are sent to readers. Sending such attachments via e-mail or with a memo lacks the formality and respect readers deserve.

7. A letter is still the most formal and approved way to conduct business with many international audiences. These readers see a letter as more polite and honorable than an e-mail for initial contacts and even for subsequent business communications.

Letter Formats

Letter format refers to the way in which you print a letter—where you indent and where you place certain kinds of information. Several letter formats exist. Two of the most frequently used business letter formats are full-block and modified.

Full-Block Format

In the full-block format all information is flush against the left margin, with spaces between paragraphs. Figure 4.1 (p. 84) shows a full-block letter. Use this format only when your letter is on **letterhead stationery** (specially printed giving a company's name and logo, business and Web addresses, fax and telephone numbers, and sometimes the names of executives).

Modified Format

The modified style (Figure 4.2, p. 85) positions the writer's address (if it is not imprinted on a letterhead), date, complimentary close, and signature to the right side of the letter. The date aligns with the complimentary close, and notations of any enclosures with the letter are flush left below the signature. Paragraphs in the modified style can be indented as in the figure, or not.

Continuing Pages

To indicate subsequent pages if your letter runs beyond one page, use one of these conventions. Note the use of the recipient's name.

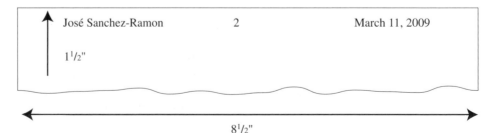

2
A.D. Smith
January 15, 2009

Guidelines on Printing Your Letters

Here are some tips on printing professional-looking letters:

- Use a letter-quality printer and check toner cartridge levels to avoid sending a fuzzy, faint, or messy letter.
- Avoid script or other fancy type fonts. See pages 174–176.
- Leave generous margins of an inch and a half all around your message.
- Single-space within each paragraph but double-space between paragraphs.
- Be careful about lopsided letters. Don't start a brief letter at the top of the page and leave the lower three-fourths of the page blank; similarly, start a shorter letter near the center of the page.
- Don't cram a long letter onto one page by squeezing too many characters on a line. Go to a second page, as in Figure 4.7 (pp. 98–100). Use the print preview command to view an image of your letter.
- Use high-quality, white bond paper.

Parts of a Letter

A letter contains many parts, each of which contributes to your overall message. The parts and their placement in your letter form the basic conventions of effective letter writing. Readers look for certain information in key places.

In the following sections, those parts of a letter marked with an asterisk should appear in every letter you write. Figure 4.3 (p. 86) is a sample letter containing all the parts discussed here. Note where each part is placed in the letter.

*Date Line

Spell out the name of the month in full—"September" or "March" rather than "Sept." or "Mar." The date line is usually keyboarded this way: November 12, 2008.

*Inside Address

The inside address, the same address as you put on the envelope, is always placed against the left margin, two lines below the date line. It contains the name, title (if any), company, street address, city, state, and ZIP code of the person to whom you are writing. Single-space the inside address and do not use any punctuation at the end of the lines.

Figure 4.1   Full block letter format.

NIRA   Nevada Insurance Research Agency
7500 South Maplewood Drive, Las Vegas, NV 89152-0026
(702) 555-9876 http://www.NIRA.org

All printing lined up against the left-hand margin

April 6, 2009

Ms. Molly Georgopolous, C.P.A.
Business Manager
Meyers, Inc.
3400 South Madison Road
Reno, NV 89554-3212

Dear Ms. Georgopolous:

Professional, businesslike font choice

As I promised in our telephone conversation earlier this afternoon, I am enclosing a study of the Nevada financial responsibility law. I hope that it will help you prepare your report.

Text of letter balanced on the page

Let me emphasize again that probably 95 percent of all individuals who are involved in an accident obtain reimbursement for medical bills and for damages to their automobiles. If individuals have insurance, they can receive reimbursement from their own carrier. If they do not have insurance and the other driver is uninsured and judged to be at fault, the Nevada Bureau of Motor Vehicles revokes that party's driver's license until all costs and damages are paid.

Paragraphs single-spaced, with an additional space between paragraphs

Please call me again if I can help you.

Sincerely yours,

Carmen Tredeau

Signature written in ink

Carmen Tredeau, President

Encl.

Generous margins on all sides of the letter

Bradley Fuller, CPCU	Carmen Tredeau, CPCU	Theodore Kendrick	Iping Li, CPCU	Dora Salinas-Diego, CPCU
Chairperson	President	Vice President Public Affairs	Vice President Research	Vice President Actuary

Modified letter format.   Figure 4.2

Arthur T. McCormack

7239 East Daphne Parkway, Mobile, AL   36608-1012

September 30, 2009

Mr. Travis Boykin, Manager
Scandia Gifts
703 Hardy St.
Hattiesburg, MS 39401-4633

Dear Mr. Boykin:

I would appreciate knowing if you currently stock the Crescent
pattern of model 5678 and how much you charge per model
number. I would also like to know if you have special prices per
box order.

The name of your store is listed on the Crescent website as the
closest distributor of Copenhagen products in my area. Would you
please give me directions to your shop from Mobile and the hours
you are open?

I look forward to hearing from you.

Sincerely yours,

Arthur T. McCormack

*Date is
indented.*

*Inside address
is single-spaced*

*Paragraphs
can be
indented or
not indented.*

*Complimentary
close and
writer's name
are indented.*

Figure 4.3   A sample letter, full block format, with all parts labeled.

Letterhead

M& **Madison and Moore, Inc.**
Professional Architects

7900 South Manheim Road
Crystal Springs, NE 71003-0092
Phone 402-555-2300 **http://www.MMI.com**

Date line

July 12, 2009

Inside address

Ms. Paula Jordan
Systems Consultant
Broadacres Development Corp.
12 East River Street
Detroit, MI 48001-0422

Salutation

Dear Ms. Jordan:

Thank you for your letter of July 6, 2009. I have discussed your request
with the staff in our planning department and have learned that the design
modules we used for our Vestuvia project are no longer available.

Body of letter

In searching through my files, however, I have come across the enclosed
catalog from a California firm that might be helpful to you. This firm,
California Concepts, offers plans very similar to the ones you are interested
in, as you can tell from the design I checked on page 23 of their catalog.

I hope this will help you and I wish you every success in your project.

*Complimentary
close*

Sincerely yours,

*Company
name*

MADISON AND MOORE, INC.

Signature

William Newhouse

*Writer's name
and title*

William Newhouse
Office Manager

*Enclosure
Copy to*

Encl.: Catalog
cc: Planning Department

Dr. Mary Petro
Director of Research
Midwest Laboratories
1700 Oak Drive
Rapid City, SD 56213-3406

Always try to write to a specific person rather than just "Sales Manager" or "President." To find out the person's name, check previous correspondence, e-mail lists, the company's or individual's website, or call the company. Make sure you use an abbreviated courtesy title (Ms., Mr., Dr., Prof.) before the recipient's name for the inside address (e.g., Capt. María Torres; Mr. A. T. Ricks; Rev. Siam Tau). Use Ms. when writing to a woman unless she has expressly asked to be called Miss or Mrs.

The last line of the inside address contains the city, state, and ZIP code.

*Salutation

Begin with *Dear,* and then follow with a courtesy title, the reader's last name, and a colon (Dear Mr. Brown:). **Never use a comma in the salutation of a formal letter.** Never use the sexist "Dear Sir," "Gentlemen," or "Dear Madam," and avoid the stilted "Ladies and Gentlemen" or "Dear Sir/Madam." (For a discussion of sexist language and how to avoid it, see Chapter 2, pp. 43–45.)

Sometimes you may not be sure of the sex of the reader. There are women named Stacy, Robin, and Lee, and men named Leslie, Kim, and Kelly. If you aren't certain, you can use the reader's full name: "Dear Terry Banks." If you know the person's title, you might write "Dear Credit Manager Banks."

*Body of the Letter

The body of the letter contains your message (see p. 89). Some of your letters will be only a few lines long, while others may extend to three or more paragraphs. Keep your sentences short and try to hold your paragraphs to under six or seven lines.

*Complimentary Close

For most business correspondence, use one of these standard closes:

Sincerely,

Respectfully,

Sincerely yours,

Yours sincerely,

If you and your reader know each other well, you can use

Cordially,

Warmest regards,

Best wishes,

Regards,

But avoid flowery closes, such as

> Forever yours,
>
> Devotedly yours,
>
> Faithfully yours,
>
> Admiringly yours,

These belong in a romance novel, not in a business letter.

*Signature

Allow four spaces so that your signature will not look squeezed in. Always sign your name in ink. An unsigned letter indicates carelessness or, worse, indifference toward your reader. A stamped signature tells readers you could not give them personal attention

Some firms prefer using their company name along with the employee's name in the signature section. If so, type the company name in capital letters two line spaces below the complimentary close and then sign your name. Add your title underneath your typed name. Here is an example:

> Sincerely yours,
>
> THE FINELLI COMPANY
>
> *Helen Stravopoulos*
>
> Helen Stravopoulos
> Cover Coordinator

*Enclosure(s) Line

The enclosure line informs the reader that additional materials (such as brochures, diagrams, forms, contract(s), a proposal) accompany your letter.

> Enclosure
> Enclosures (2)
> Encl.: 2008 Sales Report

*Copy Notation

The abbreviation *cc:* (carbon copy) or *pc:* (photocopy) informs your reader that a copy of your letter has been sent to one or more individuals.

> cc: Service Dept.
>
> cc: Jen Lu
>   Ivor Vas

Letters are copied and sent to third parties for two reasons: (a) to document a paper trail and (b) to indicate to other readers who else is involved. Professional courtesy dictates that you tell the reader if others will receive a copy of your letter.

Organizing a Standard Business Letter

A standard business letter can be divided into an introduction, a body, and a conclusion, each section responding to or clarifying a specific issue for your recipient. These three sections can each be one paragraph long, as in Figures 4.1, 4.2, and 4.3, or the body of your letter may be two or more paragraphs.

To help readers grasp your message clearly and concisely, follow this simple plan for organizing your business letters:

- In your first paragraph tell readers why you are writing and why your letter is important to them. Acknowledge any relevant previous meetings, correspondence, or telephone calls early in the paragraph (as in Figures 4.1 and 4.3).
- Put the most significant point of each paragraph first to make it easier for the reader to find. Never bury important ideas in the middle or at the end of your paragraph.
- In the second (or subsequent) paragraph, develop the body of your message with factual support, the key details and descriptions readers need. For instance, note how in Figure 4.1 Carmen Tredeau refers to specific insurance information her reader requested.
- In your last paragraph, thank readers and be very clear and precise about what you want them to do or what you will do for them. You can let them know what will happen next, what they can expect, or any combination of these messages. Don't leave your readers hanging. End cordially and professionally.

Making a Good Impression on Your Reader

You have just learned about formatting and organizing your letters. Now we turn to the content of your letter—what you say and how you say it. Writing letters means communicating to influence your readers, not to alienate or antagonize them. Keep in mind that writers of effective letters are like successful diplomats in that they represent both their company and themselves. You want readers to see you as courteous, credible, and professional.

First, put yourself in the reader's position. What kinds of letters do you like to receive: vague, impersonal, sarcastic, pushy, or condescending; or polite, businesslike, and considerate? If you have questions, you want them answered honestly, courteously, and fully.

To send such effective letters adopt the **you attitude**; in other words, signal to readers that they and their needs are of utmost importance. Incorporating the "you attitude" means you should be able to answer "Yes" to these two questions:

1. Will my readers receive a positive image of me?
2. Have I chosen words that convey both my respect for the readers and my concern for their questions and comments?

Figures 4.4 (p. 90) and 4.5 (p. 91) contain two versions of the same letter. Which one would you rather receive?

Figure 4.4   A letter lacking the "you attitude."

Brown County • Office of the Tax Assessor

County Building, Room 200, Ventura, Missouri 56780-0101

712-555-3000

February 5, 2009

Mr. Ted Ladner
451 West Hawthorne Lane
Morris, MO 64507-3005

Dear Mr. Ladner:

You have written to the wrong office here at the County Building. There is no way we can attempt to verify the kinds of details you are demanding from Brown County.

Simply put, by carefully examining the 2008 tax bill you said you received, you should have realized that it is the Tax Collector's Office, not the Tax Assessor's, that will have to handle the problem you claim exists.

In short, call or write the Tax Collector of Brown County.

Thank you!

Tracey Kowalski

Tracey Kowalski

http://www.browncounty.gov

Tone is sarcastic and uncooperative

Use of "you" alone does not signal a positive image of reader

Insulting and curt ending and close

A you-centered revision of Figure 4.4.   **Figure 4.5**

Brown County · Office of the Tax Assessor

County Building, Room 200, Ventura, Missouri 56780-0101

712-555-3000

February 5, 2009

Mr. Ted Ladner
451 West Hawthorne Lane
Morris, MO 64507-3005

Dear Mr. Ladner:

Thank you for writing about the difficulties you encountered with your 2008 tax bill. I wish I could help you, but it is the Tax Collector's Office that issues your annual property tax bill. Our office does not prepare individual homeowners' bills.

Thanks reader and gives polite explanation

If you will kindly direct your questions to Paulette Sutton at the Brown County Tax Collector's Office, County Building, Room 100, Ventura, Missouri 56780-0100, I am sure that she will be able to assist you. Should you wish to call her, the number is 458-3455, extension 212. Her email address is psutton@bctc.gov.

Helps reader solve problem with specific information

Respectfully,

Tracey Kowalski

Tracey Kowalski

Uses appropriate close

http://www.browncounty.gov

Guidelines for Achieving the "You Attitude"

As you draft and revise your work, pay special attention to the following four guidelines for making a good impression on your reader.

1. Never forget that your reader is a real person. Avoid writing cold, impersonal letters that sound as if they were form letters or voice mail instructions. Let the readers know that you are writing to them as individuals. This letter violates every rule of personal and personable communications.

> It has come to our attention that policy number 342q-765r has been delinquent in payment and is in arrears for the sum of $302.35. To keep the policy in force for the duration of its life, a minimum payment of $50.00 must reach this office by the last day of the month. Failure to submit payment will result in the cancellation of the aforementioned policy.

The example displays no sense of one human being writing to another, of a customer with a name, personal history, or specific needs. Revised, this letter contains the necessary personal (and human) touch.

> We have not yet received your payment for your insurance policy (342q-765r). By sending us your check for $50.00 within the next two weeks, you will keep your policy in force and can continue to enjoy the financial benefits and emotional security it offers you.

The benefits to an individual reader are stressed, and the reader is addressed directly as a valued customer.

Don't be afraid of using "you" in letters. Readers will feel more friendly toward you and your message. Of course, no amount of "you's" will help if they appear in a condescending context, such as the letter in Figure 4.4.

2. Keep the reader in the forefront of your letter. Make sure the reader's needs control the tone, message, and organization of your letter—the essence of the "you attitude." Stress the "you," not the "I" or the "we." Below is a paragraph from a letter that forgets about the reader.

I-Centered Draft

```
I think that our rug shampooer is the best on the market.
Our firm has invested a lot of time and money to ensure that
it is the most economical and efficient shampooer available
today. We have found that our customers are very satisfied
with the results of our machine. We have sold thousands of
these shampooers, and we are proud of our accomplishment.
We hope that we can sell you one of our fantastic machines.
```

The draft spends its time on the machine, the company, and its sales success. Readers are interested in how *they* can benefit from the machine, not in how much profit the company makes from selling it.

To win the readers' confidence, the writer needs to show how and why they will find the product useful, economical, and worthwhile at home or at work. Here is a reader-centered revision.

You-Centered Revision

```
Our rug shampooer would make cleaning your Comfort Rest Motel
rooms easier for you. It is equipped with a heavy-duty motor
that will handle your 200 rooms with ease. Moreover, that
motor will give frequently used areas, such as the lobby or
hallways, the fresh and clean look you want for your motel.
```

3. Be courteous and tactful. Refrain from turning your letter into a punch through the mail. Don't inflame your letter or e-mail readers, as the example in Figure 3.6 shows. Again, when you capture the reader's goodwill, your rewards will be greater. The following words can leave a bad taste in the reader's mouth.

it's defective	unprofessional (job, attitude, etc.)
I insist	your failure
we reject	you contend
that's no excuse for	you should have known
totally unacceptable	your outlandish claim

Compare the following discourteous sentences with the courteous revisions.

Discourteous	Courteous Revision
We must discontinue your service unless payment is received by the date shown.	Please send us your payment by November 4 so your service will not be interrupted.
You are sorely mistaken about the contract.	We are sorry to learn about the difficulty you experienced over the service terms in our contract.
The new notebook you sold me is third-rate and you charged first-rate prices.	Since the notebook is still under warranty, I hope you can make the repairs easily and quickly.
It goes without saying that your suggestion is not worth considering.	It was thoughtful of you to send me your suggestion, but unfortunately we are unable to implement it right now.

4. Don't sound pompous or bureaucratic. Write to your reader as if you were carrying on a professional conversation with her or him. A business letter should be upbeat, simple, and to the point. It needs to be reader-friendly and believable, not stuffy and overbearing.

Don't resort to using phrases that remind readers of "legalese"—language that smells of contracts, deeds, and stuffy rooms. In the following list, the words and phrases on the left are pompous expressions that have crept into letters for years; the ones on the right are contemporary equivalents.

Pompous	Contemporary
aforementioned	previously mentioned
as per your request	as you requested
at this present writing	now
I am in receipt of	I have

attached herewith	enclosed
I am cognizant of	I know
be advised that	for your information
endeavor	try
forthwith	at once
henceforth	after this
herewith, heretofore, hereby	(drop these three "h's" entirely)
immediate future	soon
in lieu of	instead of
pursuant	concerning
remittance	payment
your letter arrived and I have same	I have your letter
under separate cover	I'm also sending you
this writer	I
we regret to inform you that	we are sorry that

Types of Business Letters

The following section discusses the common types of business correspondence that you will be expected to write on the job.

1. Inquiry letters
2. Special request letters
3. Sales letters
4. Customer relations letters
   a. Follow-up letters
   b. Complaint letters
   c. Adjustment letters
   d. Collection letters

These letter types involve a variety of formats, writing strategies, and techniques. Business letters can be classified as **positive, neutral,** or **negative,** depending on their message and the anticipated reactions of your audience. Inquiry and special request letters are examples of neutral routine letters. Letters can be positive or negative, depending on your message.

- Neutral letters request specific information about a product or service, place an order, or respond to some action or question.
- Sales letters promoting a product carry good news, according to the companies that spend millions of dollars a year preparing them.
- Customer relations letters can be positive (responding favorably to a writer's request or complaint) or negative (refusing a request, saying no to an adjustment, denying credit, seeking payment, critiquing poor performance, or announcing a product recall).

Inquiry Letters

An inquiry letter asks for information about a product, service, or procedure. Businesses frequently exchange such letters. As a customer, you too have occasion to ask in a letter about a service or a special line of products, the price, the size, the color, and delivery arrangements. The clearer your letter, the quicker and more helpful your answers are likely to be.

Figure 4.6 illustrates a letter of inquiry from Michael Ortega to a real estate office managing a large number of apartment complexes. Note that it follows these five rules for an effective inquiry letter:

- states exactly what information the writer wants
- indicates clearly why the writer must have the information
- keeps questions short and to the point
- specifies when the writer must have the information
- thanks the reader

Had Michael Ortega simply written the following very brief letter to Hillside Properties, he would not have received information he needed about size, location, and price of apartments: "Please send me some information on housing in Roanoke. My family and I plan to move there soon."

Special Request Letters

Special request letters make a special demand, not a routine inquiry. For example, these letters can ask a company for information that you as a student will use in a paper, an individual for a copy of an article or a speech, or an agency for facts that your company needs to prepare a proposal or sell a product. The person or company being asked for help stands to gain no financial reward for supplying the information; the only reward is the goodwill a response creates.

Make your request clear and easy to answer. Supply readers with an addressed, postage-paid envelope, an e-mail address, and fax and telephone numbers in case they have questions. But don't ask a company to fax a long document to you. It is discourteous to expect someone else to pay the fax charges for something you need. If you request information via e-mail, don't expect your reader to tie up computer lines by sending a lengthy attachment.

Follow these seven points when asking for information in a special request letter.

1. Make sure you address your letter to the appropriate person.
2. State who you are and why you are writing—student doing a paper, employee compiling information for a report, and so on.
3. Indicate clearly your reason for requesting the information. Mention any individuals who may have suggested you write for help and information.
4. State precisely and succinctly the questions you want answered; list, number, and separate your questions.
5. Specify exactly when you need the information. Allow sufficient time—at least three weeks. Be reasonable; don't ask for the impossible.

Figure 4.6   A letter of inquiry.

Michael Ortega
403 South Main Street Kingsport, TN 37721-0217
mortega@erols.com

April 7, 2009

Mr. Fred Stonehill
Property Manager
Hillside Properties
701 South Arbor St.
Roanoke, VA 24015-1100

Dear Mr. Stonehill:

States precise request

Please let me know if you will have any two-bedroom furnished apartments available for rent during the months of June, July, and August. I am willing to pay up to $750 a month plus utilities. My wife, one-year-old son, and I will be moving to Roanoke for the summer so I can take classes at Virginia Western Community College.

Explains need for information

Identifies area of interest

If possible, we would like to have an apartment that is within two or three miles of the college. We do not have any pets.

Specifies exact date when a reply is needed

I would appreciate hearing from you within the next two weeks. My e-mail address is **mortega@erols.com**, or you can call me at home (606-555-8957) any evening from 6–10 p.m.

Offers to confer with and thanks the reader

If you have any suitable vacancies, we would be happy to drive to Roanoke to look at them and give you a deposit to hold an apartment. Thanks for your help.

Sincerely yours,

Michael Ortega

Michael Ortega

6. Offer to forward a copy of your report, paper, or survey in thanks of the anticipated help.
7. Thank the reader for helping.

Figure 4.7 gives an example of a letter that uses the guidelines.

Sales Letters

A sales letter is written to persuade the reader to buy a product, try a service, support some cause, or participate in some activity. No matter what profession you have chosen, there will always be times you have to sell a product, a service, a community or charitable program, a point of view, or yourself!

The Four A's of Sales Letters

Successful sales letters follow a time-honored and workable plan—what can be called the "Four A's":

1. It gets the reader's *attention*—with a question or a how-to statement (e.g., We can show you how to save $1,000 on your next credit card purchase).

2. It highlights the product's or service's *appeal*—emotionally or financially, or both.

3. It shows the customer the product's or service's *application*—descriptions, special features, guarantees.

4. It ends with a specific request for *action*—call, visit, participate, register online.

These four goals can be achieved in fewer than four or five paragraphs. Look at the sales letter in Figure 4.8 (p. 100) in which these parts are labeled. Figure 4.22 (pp. 124–125) also contains a model sales letter.

Do I Mention Costs?

As a general rule, do not bluntly state the cost. Relate prices, charges, or fees to the benefits provided by the services or products to which they apply. Let customers see how much they are getting for their money, as Cory Soufas does in paragraph 3 in Figure 4.8. Similarly, a dealer who installs steel shutters did not tell readers the exact price of the product but stressed in her sales letter that they will save money by buying it: "Virtually maintenance free, your Reel Shutters also offer substantial savings in energy costs by reducing your heat loss through radiation by as much as 65% . . . and that lowers your utility bills by at least 35%."

Customer Relations Letters

Much business correspondence deals explicitly with establishing and maintaining friendly working relations. Such correspondence, known as **customer relations letters,** sends readers good news or bad news, acceptances or refusals. Good news tells customers one or more of the following:

Figure 4.7   A special request letter.

1505 West 19th Street
Syracuse, NY 13206

phone 315-555-1214

October 5, 2009

Ms. Sharonda Aimes-Worthington
Research Director
Creative Marketing Associates
198 Madison Ave.
New York, NY 10016-0092

Dear Ms. Aimes-Worthington:

Explains reason for letter

I am a junior at Monroe College in Syracuse, and am writing a report on "Internet Marketing Strategies for the Finger Lakes Region of New York" for my Marketing 340 class. Several of my professors have spoken very highly of Creative Marketing Associates, and in my own research I have learned a great deal from reading your article on Web designs and local economies posted on your website two months ago.

Proves writer has done research

Acknowledges reader's expertise

Given your extensive experience in using the Internet to promote regional businesses and tourism, I would be grateful if you would share your responses to the following three questions with me:

1. What have been the most effective ways to design websites for a regional marketplace such as the Finger Lakes?

Lists specific numbered questions on the topic

2. How can area chambers of commerce best provide links to benefit both the local municipalities and businesses in the Finger Lakes area?

3. Which other regional area(s) do you see having the same or very similar marketing goals and challenges as the Finger Lakes?

jkawatsu@webnet.com

Continued

(Continued)   **Figure 4.7**

Page 2

Your answers to these questions would make my report much more authoritative and useful. I would be happy to send you a copy and will, of course, cite you and Creative Marketing Associates in my work.

As an incentive, offers to share report

Because my report is due by December 2, I would greatly appreciate having your answers within the next month so that I can include them. Would you kindly send your responses, or any questions you may have, to my e-mail address, *jkawatsu@webnet.com.* Many thanks for your help.

Indicates when needed

Makes contact easy through e-mail

Sincerely yours,

Julie Kawatsu

Julie Kawatsu

- You agree with them about a problem they brought to your attention.
- You are solving their problem exactly the way they want.
- You are approving their loan or request for a refund.
- You are grateful to them for their business.

Thank you letters, congratulations letters, and adjustment letters saying "Yes" with these messages are all examples of good news messages.

Bad news messages, however, inform readers that:

- You do not like their work or the equipment they sold you.
- You do not have the equipment or service they want or you cannot provide it at the price they want to pay.
- You cannot refund their purchase price or perform a service.
- You are raising their rent or not renewing their lease.
- You want them to pay what they owe you now.

Bad news messages often come to readers through complaint letters, adjustment letters that say "No," and collection letters.

Being Direct or Indirect

Not every customer relations letter starts by giving the reader the writer's main point, judgment, conclusion, or reaction. Whether you are sending good news or bad news, determine what to say and where. *Where you place your main idea is determined by the type of letter you are writing.* Good news messages require one tactic; bad news ones, another.

Figure 4.8   A sales letter sent to a business reader.

Workwell Software
3700 Stewart Avenue   Chicago IL 60637-2210
Phone: (312) 555-3720   **Fax:** (312) 555-7601   **E-mail:** sales@workwell.com
http://www.workwell.com

August 12, 2009

Ali Jen
Circuit Systems
7 Tyler Place
Oklahoma City, OK 73101

Dear Ali Jen:

Gets reader's attention with a question

Do you know how much money your company loses from repetitive strain injury? Each year employers spend millions of dollars on employee insurance claims because of back pains, fatigue, eye strain, and carpal tunnel syndrome.

Emphasizes the product's appeal

Workwell can solve your problems with its easy-to-use Exercise Program Software that will automatically monitor the time employees spend at their computers and also measure keyboard activity. After each hour (or the specified number of keystrokes), **Workwell** software will take your employees through a series of exercises that will help prevent carpal tunnel syndrome and strains.

Shows specific application of the product

Links costs to benefits

Workwell's Exercise Program Software will not interfere with busy schedules. Each of the 27 exercises is demonstrated on screen with audio instructions. The entire program takes less than 3 minutes and is available for Windows XP and Windows Vista. For only $1499.00, you can provide a networked version of this valuable software to all of your employees.

Ends with a call for action

To help your employees stay at peak efficiency in a safe work environment, please call us at 1-800-555-WELL or contact us at **http://www.workwell.com** to order your software today.

Thank you,

Cory Soufas
Sales Manager

Good News Message

If you are writing a good news letter, use the direct approach. Start your letter with the welcome, pleasant news that the reader wants to hear. Don't postpone the opportunity to put your reader in the right frame of mind. Then provide any relevant supporting details, explanations, or commentary. Being direct is advantageous when you have good news to convey.

Bad News Message

If you have bad news to report, do *not* open your letter with it. Be indirect. Prepare your reader for the bad news; keep the tension level down. If you throw the bad news at your reader right away, you jeopardize the goodwill you want to create and sustain. Consider how you would react to a letter that begins with these slaps:

- Your order cannot be filled.
- Your application for a loan has been denied.
- It is our unfortunate duty to report . . .

Having been denied, disappointed, or even offended in the first sentence or paragraph, the reader is not likely to give you his or her attentive cooperation thereafter.

Two Versions of a Bad News Message   Notice how A. J. Griffin's bad news letter in Figure 4.9 curtly starts off with the bad news of a rent increase. Receiving such a letter, the owner of Flowers by Dan certainly could not be blamed for looking for a new location for his business.

Compare the curt version of Griffin's letter in Figure 4.9 with his revised message in Figure 4.10 (p. 103). In the revised version, Griffin begins tactfully with pleasant, positive words designed to put his reader in a good frame of mind. Then Griffin gives some background information that the owner of Flowers by Dan can relate to. Griffin makes one more attempt to encourage Sobol to recall his good feelings about the Mall—last year they did not raise rents—before introducing the bad news of a rent increase.

Even after giving the bad news, Griffin softens the blow by saying that the Mall knows it is bad news. Griffin's tactic here is to defuse some of the anger that Sobol will inevitably feel. Griffin then ends on a positive, upbeat note: a prosperous future for Flowers by Dan.

Follow-Up Letters

A follow-up letter is sent by a company after a sale to thank the customer for buying a product or using a service and to encourage the customer to buy more products and services. A follow-up letter is a combination thank-you note and sales letter. The letter in Figure 4.11 (p. 105) shows how an income tax preparation service attempts to obtain repeat business by doing the following:

1. begins with a brief and sincere expression of gratitude
2. discusses the benefits (advantages) the customer already knows about and then transfers the firm's dedication to the customer to a continuing sales area
3. ends with a specific request for future business

Figure 4.9   An ineffective bad news letter.

River Road Mall

December 1, 2008

Mr. Daniel Sobol
Flowers by Dan
Lower Level 107
River Road Mall

Dear Mr. Sobol:

Blunt opening disregards audience's needs and feelings

This is to inform you of a rent increase. Starting next month your new rent will be $3,500.00, resulting in a 15 percent increase.

Ends with a demand

Please make sure that your January rent check includes this increase.

Sincerely,

A. J. Griffin
Manager
ajg@rrmall.com

300 First Street
Canton, Ohio 44701
(216) 555-6700
www.RRMall.com

A diplomatic revision of the bad news letter in Figure 4.9. **Figure 4.10**

River Road Mall

December 1, 2008

Mr. Daniel Sobol
Flowers by Dan
Lower Level 107
River Road Mall

Dear Mr. Sobol:

It has been a pleasure to have you as a tenant at the Mall for the past two years, and we look forward to serving you in the future.

Opens with positive association

Over these last two years we have experienced a dramatic increase in costs at River Road Mall for security, maintenance, landscaping, pest control, utilities, insurance, and taxes. Last year we absorbed those increases and so did not have to raise your rent. We wish we could do it again but, unfortunately, we must increase your rent by 15 percent, to $3,500.00 a month, effective January 1.

Prepares reader for bad news to follow

States bad news in most concise, up-beat way

Although no one likes a rent increase, we know that you do not want us to compromise on the quality of service that you and your customers expect and deserve here at River Road Mall.

Links bad news to reader benefits

Please let us know how we can assist you in the future. We wish you a very successful and profitable 2009. If you have any questions, please call or visit my office.

Does not apologize but ends with respectful tone

Cordially,

A. J. Griffin

A. J. Griffin
Manager
ajg@rrmall.com

Warm, complimentary close

300 First Street
Canton, Ohio 44701
(216) 555-6700
www.RRMall.com

Complaint Letters

Each of us, either as customers or businesspeople, at some time has been frustrated by a defective product, inadequate or rude service, or incorrect billing. When we get no satisfaction from calling an 800 number and are routed through a series of menu options, our frustration level goes up. Usually our first response is to write a letter dripping with juicy insults. But an angry letter, like a piece of flaming e-mail (see Figure 3.6, p. 76), rarely gets positive results and can hurt your company's image.

A complaint letter is a delicate one to write. First off, avoid the following:

- name calling
- sarcasm
- insults
- threats
- unflattering clip art
- intimidating type fonts (e.g., all capital letters)

The key thing to keep in mind is that you can disagree without being disagreeable. Be rational, not hostile. Just to let off steam, you might want to write an angry letter but then tear it up, replacing all the heat with courteous and diplomatic language.

Establishing the Right Tone

A complaint letter is written for more reasons than just blowing off steam. You want some specific action taken. The "you attitude" is especially important here to maintain the reader's goodwill. An effective complaint letter can be written by an individual consumer or by a company. Figure 4.12 (p. 106) shows Michael Trigg's complaint about a defective fishing reel; Figure 4.13 (p. 107) expresses a restaurant's dissatisfaction with an industrial dishwasher.

Writing an Effective Complaint Letter

To increase your chances of receiving a speedy settlement, follow these five steps in writing your letter of complaint. They will help you build your case.

1. Begin with a detailed description of the product or service. Give the appropriate model and serial numbers, size, quantity, color, and cost. Specify check and invoice numbers. Indicate when, where (specific address), and how (through a vendor, the Internet, at a store) you purchased it and also the remaining warranty. If you are complaining about a service, give the name of the company, the date of the service, the personnel providing it, and their exact duties.

2. State exactly what is wrong with the product or service. Precise information will enable the reader to understand and act on your complaint.

- How many times did the product work before it stopped?
- What parts were malfunctioning?
- What parts of a job were not done or were done poorly?
- When did all this happen? How many times?
- Where and how were you inconvenienced?

A follow-up letter to encourage repeat business.   **Figure 4.11**

Taylor Tax Service
Highway 10
North Jennings, TX 78326
phone (888) 555-9681 e-mail taylor@aol.com
http://www.taylor.com

December 2, 2009

Ms. Laurie Pavlovich
345 Jefferson St.
Jennings, TX 78326

Dear Ms. Pavlovich:

Thank you for using our services in February of this year. We were pleased to help you prepare your 2008 federal and state income tax returns. Our goal is to save you every tax dollar to which you are entitled. If you ever have questions about your return, we are open all year long to help you.

Links business goal to customer advantage

We are looking forward to serving you again next year. Several new federal tax laws will change the types of deductions you can declare. These changes might appreciably increase your refund. Our consultants know the new laws and are ready to apply them to your return.

Stresses reasons/benefits for customer to return for service

Another important tax matter influencing your 2008 returns will be any losses you may have suffered because of the hailstorms and tornadoes that hit our area three months ago. Our consultants are specially trained to assist you in filing proper damage claims with your federal and state returns.

Makes it easy and profitable for customer to act soon

To make using our services even more convenient for you, we will file your tax return electronically to speed up any refund to which you are entitled. Please call us at (888) 555-9681 or e-mail us at **taylor@aol.com** as soon as you have received all your forms to set up an appointment. We are waiting to serve you seven days a week from 9:00 a.m. to 9:00 p.m.

Ends with commitment to customer convenience

Sincerely yours,

TAYLOR TAX SERVICE

Demetria Taylor

Demetria Taylor

Figure 4.12   A complaint letter from a consumer.

17 Westwood Drive

Magnolia, MA 02171

mtrigg@roof.com

October 10, 2008

Mr. Ralph Montoya
Customer Relations Department
Smith Sports Equipment
P.O. Box 1014
Tulsa, OK 74109-1014

Dear Mr. Montoya:

Documents all relevant details about the product

On September 21, 2008, I purchased a Smith reel, model 191, at the Uni-Mart Store on Marsh Avenue in Magnolia, Massachusetts. The reel sold for $84.95 plus tax. The reel is not working effectively, and I am returning it to you under separate cover by first-class mail.

Explains politely what is wrong

I had made no more than five casts with the reel when it began to malfunction. The button that releases the spool and allows the line to cast will not spring back into position after casting. In addition, the gears make a grinding noise when I try to retrieve the line. Because of these problems, I was unable to continue my participation in the Gloucester Fishing Tournament last week.

Clearly states what should be done

I am requesting that a new reel be sent to me free of charge in place of the defective one I returned. I would also like to know what was wrong with the defective reel.

Specifies an acceptable time frame

I would appreciate your processing my claim within the next two weeks.

Sincerely yours,

Michael Trigg

Michael Trigg

A complaint letter from a business.   Figure 4.13

The Loft

Camerson and Dale, Sunnyside, California 91793-4116   213-555-7500

June 21, 2009

Priscilla Dubrow
Customer Relations Department
Superflex Products
San Diego, CA 93141-0808

Dear Ms. Dubrow:

On September 15, 2008, we purchased a Superflex industrial dishwasher, model 3203876, at the Hillcrest store at 3400 Broadway Drive in Sunnyside, for $5,000. In the last three weeks, our restaurant has had repeated problems with this machine. Three more months of warranty remain on the unit.

The machine does not complete a full cycle; it stops before the final rinsing and thus leaves the dishes dirty. It appears that the cycle regulators are not working properly because they refuse to shift into the next necessary gear. Attempts to repair the machine by the Hillcrest service team on June 4, 11, and 15 have been unsuccessful.

The Loft has been greatly inconvenienced. Our kitchen team has been forced to sort, clean, and sanitize utensils, dishes, pans and pots by hand, resulting in additional overtime. Moreover, our expenses for proper detergents have increased.

We want your main office to send another service team at once to fix this machine. If your team is unable to do this, we want a discount worth the amount of the warranty life on this model to be applied to the purchase of a new Superflex dishwasher. This amount would come to $1,000, or 20 percent of the original purchase price.

So that our business is not further disrupted, we would appreciate your resolving this problem promptly within the next week.

Sincerely yours,

Emily Rashon

Emily Rashon
Co-owner

▶ Browse our menu, which changes daily, at www.theloft.com

Gives all product's facts and warranty information

Describes what's happened and when

Further documents problem and reason for adjustment

Provides clear description of how problem should be solved

Concludes politely with justification for prompt action

Stating that "the brake shoes were defective" tells very little about how long they were on your car, how effectively they may have been installed, or what condition they were in when they ceased functioning safely.

3. Briefly describe the inconvenience you have experienced. Show that your problems were directly caused by the defective product or service. To build your case, give precise details about the time and money you lost. Don't just say you had "numerous difficulties." Did you have to pay a mechanic to fix your car when it was stalled on the road? Did you have to buy a new printer or DVD player? Where appropriate, refer to any previous telephone calls, e-mails, or letters. Give the names of the people you have written to or spoken with and the dates.

4. Indicate precisely what you want done. Do not simply write that you "want something done." State precisely that you want one or more of the following:

- your purchase price refunded
- your model repaired or replaced
- a completely new repair crew provided
- an apology from the company for discourteous treatment

5. Ask for prompt handling of your claim. In your concluding paragraph, ask the reader to answer any question you may have (such as finding out where calls came from that you were billed for but did not make). You might even specify a reasonable time by which you want to hear from the reader or need the problem fixed. Note the last paragraph in Figure 4.13.

Adjustment Letters

Adjustment letters respond to complaint letters by telling customers dissatisfied with a product or service how their claim will be settled. Adjustment letters should reconcile the differences that exist between a customer and a firm and restore the customer's confidence in that firm.

Adjustment Letters That Tell the Customer "Yes"

It is easy to write a "Yes" letter if you remember a few useful suggestions. As with a good news message, start with the favorable news the customer wants to hear; that will put him or her in a positive frame of mind to read the rest of your letter. Let the customer know that you sincerely agree with him or her—don't sound as if you are reluctantly honoring the request.

The two examples of adjustment letters saying "Yes" show you how to write this kind of correspondence. The first example, Figure 4.14, says "Yes" to Michael Trigg's letter in Figure 4.12. Reread the Trigg complaint letter to see what problems Ralph Montoya faced when he had to write to Mr. Trigg. The second example of an adjustment letter that says "Yes" is in Figure 4.15. It responds to a customer who has complained about an incorrect billing.

An adjustment letter saying "Yes."   **Figure 4.14**

Smith Sports Equipment
P.O. Box 1014   Tulsa, Oklahoma 74109-1014
(918) 555-0164 ▪ www.smithsport.com

October 19, 2008

Mr. Michael Trigg
17 Westwood Drive
Magnolia, MA 02171

Dear Mr. Trigg:

Thank you for alerting us in your letter of October 10 to the problems you had with one of our model 191 spincast reels. I am sorry for the inconvenience the reel caused you. A new Smith reel is on its way to you.

We have examined your reel and found the problem. It seems that a retaining pin on the button spring was improperly installed by one of our new soldering machines on the assembly line. We have thoroughly inspected, repaired, and cleaned the soldering machine to eliminate the problem from happening again.

Since we began making quality reels in 1955, we have taken pride in helping loyal customers like you who rely on a Smith reel. We hope that your new Smith reel brings you years of pleasure and many good catches, especially next year at the Gloucester Fishing Tournament.

Thank you for your business. Please let me know if I can assist you again.

Respectfully,

SMITH SPORTS EQUIPMENT

Ralph Montoya

Ralph Montoya, Manager
Customer Relations Department

Apologizes and announces good news

Explains what happened and why problem will not recur

Expresses respect for customer

Closes with friendly offer to help again

Figure 4.15   An adjustment letter saying "Yes."

Brunelli Motors

- -

Route 3A, Giddings, Kansas 62034-8100 (913) 555-1521

August 5, 2009

Ms. Kathryn Brumfield
34 East Main
Giddings, KS 62034-1123

Dear Ms. Brumfield:

Thanks customer and complies with request

We appreciate your notifying us, in your letter of July 30, about the problem you experienced regarding warranty coverage on your new Phantom Hawk GT. The bills sent to you were incorrect, and I have cancelled them. Please accept my apologies. You should not have been charged for a shroud or for repairs to the damaged fan and hose, since all those parts, and labor on them, are covered by your warranty.

Explains why problem occurred and how it has been resolved

The problem was the result of an error in the way the charges were listed. Our firm has begun using new software to give customers better service, and the mechanic apparently entered the wrong code for your account. We have since programmed our system to flag any bills for vehicles still under warranty. We hope that this new procedure will help us serve you and our other customers more efficiently.

Ends courteously and leaves reader with good feeling about the dealership

We value you as a customer of Brunelli Motors. When you are ready for another Phantom, I hope that you will once again visit our dealership.

Sincerely yours,

Susan Chee-Saafir

Susan Chee-Saafir
Service Manager

- -

Experience virtual reality: Drive a new Phantom at
http://www.brunelli.com

Writing a "Yes" Letter

The following four steps will help you write a "Yes" adjustment letter.

1. Admit immediately that the customer's complaint is justified and apologize. Briefly state that you are sorry and thank the customer for writing to inform you.

2. State precisely what you are going to do to correct the problem. Let the customer know that you will

- extend warranty coverage
- credit an account with funds, more air miles, and so on
- offer a discount on next purchase
- cancel a bill or give credit toward another purchase
- repair/replace damaged equipment
- give the customer credit toward another purchase
- upgrade software, accommodations, etc.

In Figure 4.14 Michael Trigg is told right away that he will receive a new reel; in Figure 4.15 Kathryn Brumfield learns in the first paragraph that she will not be charged for parts or service.

3. Tell customers exactly what happened. They deserve an explanation for the inconvenience they suffered. Note that the explanations in Figures 4.14 and 4.15 (pp. 109–110) give only the essential details; they do not bother the reader with side issues or petty remarks about who was to blame. Assure customers that the mishap is not typical of your company's operations.

4. End on a friendly—and positive—note. Don't remind customers about their trouble. Leave customers with a good feeling about your company.

Adjustment Letters That Tell the Customer "No"

Writing to tell customers "No" is obviously more difficult than agreeing with them. You are faced with the sensitive task of conveying bad news, while at the same time convincing the reader that your position is fair, logical, and consistent. Do not bluntly start off with a "No." Do not accuse or argue. Avoid remarks that blame, scold, or remind customers of a wrongdoing.

How to Say "No" Diplomatically

The following five suggestions will help you say "No" diplomatically. Practical applications of these suggestions can be found in Figures 4.16 (p. 112) and 4.17 (p. 113). Contrast the refusal of Michael Trigg's complaint in Figure 4.16 with the favorable response to it in Figure 4.14.

1. Thank customers for writing. Make a friendly start by putting them in a good frame of mind. Don't put readers on the defensive by starting off with "We regret to inform you." The letter writers in Figures 4.16 and 4.17 thank the customers for bringing the matter to their attention. As with other bad news letters, never begin with a refusal. You need time to calm and convince customers. Telling them "No" in the first

Figure 4.16   An adjustment letter saying "No."

Smith Sports Equipment

P.O. Box 1014   Tulsa, Oklahoma 74109-1014
(918) 555-0164 ▪ www.smithsport.com

October 19, 2008

Mr. Michael Trigg
17 Westwood Drive
Magnolia, MA 02171

Dear Mr. Trigg:

Thanks and sympathizes with reader

Thank you for writing to us on October 10 about the trouble you experienced with our model 191 spincast reel. We are sorry to hear about the difficulties you had with the release button and gears.

Explains problem without directly blaming the reader; gives firm decision

We have examined your reel and found the trouble. It seems that a retaining pin in the button spring was pushed into the side of the reel casing, thereby making the gears inoperable. The retaining pin is a vital yet delicate part of your reel. In order to function properly, it has to be pushed gently. Since our replacement guarantee does not cover the use of the pin in this way, we cannot refund your purchase price.

Turns a "No" into a "Yes" for customer

However, we want you to have many more hours of fishing pleasure and so we would be happy to repair your reel for $29.98 and return it to you within 10 days. Please let us know your decision.

Ends politely without any reference to the problem

I look forward to hearing from you.

Respectfully,

SMITH SPORTS EQUIPMENT

Ralph Montoya

Ralph Montoya, Manager
Customer Relations Department

Health AIR

4300 Marshall Drive
Salt Lake City, Utah 84113-1521
(801) 555-6028
www.healthair.com

November 20, 2009

Ms. Denise Southby, Director
Bradley General Hospital
Bradley, IL 60610-4615

Dear Director Southby:

Thank you for your letter of November 14 explaining the problems you have encountered with our Puritan MAII ventilator. We were sorry to learn that you were unable to get the high-volume PAO_2 alarm circuit to work.

Professional you-centered opening

Our ventilator is a high-volume, low-frequency machine that can deliver up to 40 ml. of water pressure. The ventilator runs with a center of gravity attachment on the right side of the diode. The trouble you had with the high oxygen alarm system is due to an overload on your piped-in oxygen. Our laboratory inspection of the ventilator you returned indicated that the high-pressure system had blown a vital adaptor in the MAII. An overload in an oxygen system is a condition that is not included under warranty coverage of the ventilator, and so we cannot replace it free of charge.

Justifies firm decision by explaining causes of problem and conditions of sale

We would, however, be pleased to send you another model of the adaptor, which would be more compatible with your system, as soon as we receive your order. The price of the adaptor is $600, and our service representative will be happy to install it for you at no charge.

Provides practical alternative with an incentive

Please let me know your decision. I look forward to hearing from you.

Ends with goodwill

Sincerely yours,

R. P. Gifford

R. P. Gifford
Customer Service Department

sentence or two will negatively color their reactions to the rest of the letter. Use the indirect approach discussed earlier in the chapter (p. 101).

2. State the problem carefully to reassure the customer that you understand the complaint. You thereby prove that you are not trying to misrepresent or distort what the customer has told you.

3. Explain what happened with the product or service before you give the customer a decision. Provide a factual explanation to show the customer that he or she is being treated fairly. Rather than focusing on the customer's misunderstanding the instructions or a failure to observe details of a service contract, state the proper ways of handling a piece of equipment.

> **Poor:** By reading the instructions on the side of the paint can, you would have avoided the streaking condition that you claim resulted.
>
> **Revised:** Hi-Gloss Paint requires two applications, four hours apart, for a clear and smooth finish.

The revision reminds the customer of the right way to apply the paint without pointing an accusing finger. Note how the explanations in Figures 4.16 and 4.17 emphasize the appropriate way of using the product.

4. Give your decision without hedging. Do not say, "Perhaps some type of restitution could be made later" or "Further proof would have been helpful." Indecision will infuriate customers who believe that they have already presented a sound, convincing case. Never apologize for your decision.

5. Leave the door open for better and continued business. Whenever possible, help customers solve their problem by offering to send them a new product or part and quote the full sales price. Note how the second-to-last paragraph in Figure 4.16 and the last paragraph in 4.17 do that diplomatically.

Collection Letters

Collection letters require the same tact and fairness as do complaint and adjustment letters. Each nonpayment case should be evaluated separately. A nasty collection letter sent to a customer who is a good credit risk after only one month's nonpayment can send the customer elsewhere. On the other hand, three easygoing letters to a customer who is a poor credit risk may encourage that individual to postpone payment, perhaps indefinitely.

Types of Collection Letters

Many businesses send several letters to customers before turning matters over to a collection agency. Each letter in the series employs a different technique, ranging from giving compliments and offering flexible credit terms to issuing demands for immediate payment and threats of legal consequences. One hospital uses the collection letters illustrated in Figures 4.18 (p. 115) and 4.19 (p. 116) to encourage patients to pay their bills. Figure 4.18 is a letter sent early in the collection process when a client is only a month or two late. The collection letter in Figure 4.19 is sent much later to a client who has ignored earlier notices.

A first, or early, collection letter.   **Figure 4.18**

SABINE COUNTY HOSPITAL
7200 Medical Blvd.
Sabine, TX 77231-0011
512-555-6734
www.sabine.org

May 15, 2009

Mr. Cal Smith
24 Mulberry Street
Valley, TX 77212-3160

Re: Inpatient Services
Date of Hospitalization: March 11–12, 2009
Balance Due: $4,725.48

Dear Mr. Smith:

We are honored that we were able to serve your health care needs during your recent stay at Sabine County Hospital. It is our continuing goal to provide the best possible hospital care for residents of Sabine County and its vicinity. To do so we must keep our finances up-to-date.

Our records indicate that your account is now overdue, and that we have not received a payment from you for two months. If you have recently sent one in, kindly disregard this letter and accept our thanks.

If for any reason you are unable to pay the full amount at this time, we will be happy to set up a payment schedule that is convenient for you. Just fill in the appropriate blanks below, and return this letter to us. That will enable us to avoid billing you on a "Past Due" basis. Thank you for your cooperation.

Sincerely,

Morris T. Jukes

Morris T. Jukes
Accounts Department

( ) I will pay $ _____ ( ) monthly ( ) quarterly on my account.

( ) Enclosed is a check for full payment in the amount of $ _____.

Signature

_____

Links hospital mission to patient's payment

Diplomatic reminder to pay now

Offers options to maintain goodwill

Figure 4.19   A final, collection letter.

SABINE COUNTY HOSPITAL
7200 Medical Blvd.
Sabine, TX 77231-0011
512-555-6734
www.sabine.org

September 21, 2009   Re: Inpatient Services
                                     Date of Hospitalization: March 11–12, 2009
                                     Balance Due: $4,725.48
Mr. Cal Smith
24 Mulberry Street
Valley, TX 77212-3160

Dear Mr. Smith:

Direct opening about history and status of account

During the past few months we have written to you several times about your balance of $4,725.48 for services you received on March 11 and 12. Your account is more than 190 days overdue, and we cannot allow any further extensions in receiving a payment from you.

Reminds patient of goodwill and requests payment

As you will recall, we have tried to help you meet your obligations by offering several options for paying your bill. You could have arranged for installment payments that would be due each month or quarter, whichever would be more convenient. Because you have not replied, we must ask for full payment now.

States final option

If we do not hear from you within ten days, we will have no alternative but to turn your account over to our collection agency, which will seriously hurt your credit rating. Neither of us would find this a welcome alternative.

Sincerely,

Morris T. Jukes

Morris T. Jukes
Accounts Department

The tone of Figure 4.18 is cordial and sincere—now is not the time to say "pay up or else." It stresses how valuable the person is and underscores how pleased the hospital is to have provided the care he needed. The last paragraph makes a request for payment, offering: (1) a flexible payment schedule and (2) an escape from the inconvenience (or embarrassment) of receiving past due notices. The bottom of the letter conveniently lists payment options available to the patient.

The late collection letter in Figure 4.19, on the other hand, points out that the time for concessions is over, and reminds the patient of all the efforts that the hospital has expended to collect its bills. Then it announces what consequences will result if he still does not pay.

International Business Correspondence

After e-mails, letters are the most frequent type of communication you are likely to have with international readers. Formal letter writing is a very important skill in the global marketplace. But, as we saw in Chapter 1, you cannot assume that every culture writes letters the way we do in the United States. The conventions of letter writing—formats, inside addresses, salutations, dates, complimentary closes, signature lines—are as diverse as the organization and styles used in writing letters for international audiences. You may also need to communicate from an international perspective with a company located in the United States. For instance, restaurant owner Patrice St. Jacques appeals to the ethnic heritage and pride of a potential customer in an effective sales letter (see Figure 4.22 on pp. 124–125).

It would be impossible to give you information about how to write letters to each international audience. There are at least 5,000 major languages representing diverse ethnic and cultural communities around the globe. But here are some of the most important culturally sensitive questions you need to ask about writing to readers whose cultures are different from yours:

- What is your status in relationship to the reader (client, vendor, salesperson, or international colleague)?
- How should you format and address your letter?
- What is an appropriate salutation?
- How should you begin and conclude your letter?
- What types and amount of information will you have to give?
- What is the most appropriate tone to use?

To answer these and similar questions about proper letter protocol for your international readers, you need to learn about their culture on the Internet, from multinational colleagues, or through foreign language instructors.

Chapter 4 Additional Activities, located at **college.cengage .com/pic/ kolinconcise2e**, include a global-focused activity titled "Writing Letters for International Readers."

Guidelines for Communicating with International Readers

When you communicate with international readers, you need to be aware of cultural differences between you and your reader. The conventions of writing—the words, sentences, even the type of information you offer—can and do change from one culture to another.

The following nine guidelines will help you communicate more successfully with an international audience, significantly reducing the chances of their misunderstanding you.

1. Use common, easily understood vocabulary. Write basic, simplified English. Choose words that are widely understood as opposed to those that are not used or comprehended by many speakers. Avoid low-frequency words by substituting simpler synonyms; for example, use *stop*, not *refrain; prevent*, not *forestall; discharge*, not *exude; happy*, not *exultant*.

2. Avoid ambiguity. Words that have double meanings force non-native readers to wonder which one you mean. For example, "We fired the engine" would baffle your readers if they were not aware of the multiple meanings of *fire*. Unfamiliar with the context in which *fire* means "start up," a non-native speaker of English might think you're referring to "setting on fire or inflaming," which is not what you intend. Such misinterpretation is likely because most bilingual dictionaries would probably list only those two meanings.

3. Be careful about technical vocabulary. While a reader who is a non-native speaker may be more familiar with technical terms than with other English words, make sure the technical word or phrase you include is widely known and not a word or meaning used only at your plant or office. Double-check by consulting the most up-to-date manuals and guides in your field, but steer clear of technical terms in fields other than the one with which your reader is familiar. Be especially careful about using business words and phrases an international reader may not know, such as "lean manufacturing," "revolving credit," "reverse mortgage," "amortization," and so forth.

4. Avoid idiomatic expressions. The following colorful idiomatic expressions will confuse and may even startle a non-native reader:

I'm all ears	think outside the box
throw cold water on it	sleep on it
hit the nail on the head	give a heads up to
new blood	land in hot water
easy come, easy go	touch and go
get a handle on it	pushed the envelope
right under your nose	it was a rough go

The meanings of those and similar phrases are not literal but figurative—a reflection of our culture, not necessarily your reader's. A non-native speaker of English will approach such phrases as combinations of the separate meanings of the individual words, not as a collective unit of meaning.

A non-native speaker of English—a potential customer in Asia or Africa, for example—might be shocked if you wrote about a sale concluded at a branch office this way: "Last week we made a killing in our office." Substitute a clear, unambiguous translation easily understood in international English: "We made a big sale last week." For "Sleep on it," you might say, "Please take a few days to make your decision."

5. Delete sports and gambling metaphors. These metaphors, which are often rooted in American popular culture, do not translate word for word for non-native speakers and so again can interfere with communication with your readers. Here are a few examples to avoid:

out in left field	a ballpark figure
struck out	fumbled the ball
dropped the ball	out of bounds
down for the count	made a pass
long shot	beat the odds

Use a basic English dictionary and your common sense to find nonfigurative translations for the preceding and similar expressions.

6. Don't use abbreviations, acronyms, or contractions. While these shortened forms of words and phrases are a part of U.S. business culture, they might easily be misunderstood by a non-native speaker. Avoid abbreviations such as pharm., gov., org., pkwy, rec., hdg., hr., mfg., or w/o. The following acronyms can also cause your international reader trouble: ASAP, PDQ, p's and q's, IRA, SUV, RV, DOB, DOT, SSN. Contractions such as the following might lead readers to mistake them for the English word they look like: I've (ivy), he'll (hell), he'ld (held), I'll (ill), we'll (well), can't (cant), won't (wont, want).

7. Watch units of measure. Do not fall into the cultural trap of assuming that your reader measures distances in miles and feet (instead of kilometers and meters as most of the world does), measures temperature on a Fahrenheit scale (instead of Celsius), buys gallons of gasoline (instead of liters), and spends dollars (rather than euros, pesos, marks, rupees, or yen). Always respect the monetary unit your reader uses. Also, many countries follow a 24-hour clock without a.m. or p.m. In Paris, for instance, 13:30 is 1:30 p.m. in New York, and 9 p.m. in the United States translates into 21:00 in Sweden. Adapt your message to the readers' culture.

8. Avoid culture-bound descriptions of place and space. For example, when you tell a reader in Hong Kong about the "Sunbelt" or a potential client in Africa about the "Big Easy," will he or she know what you mean? When you write from California to a non-native English speaker in India about the eastern seaboard, meaning the East Coast of the United States, the directional reference may not mean the same thing to your audience as it does to you. Be respectful of your readers' cultural (and physical) environment as well. Thanksgiving is celebrated in the United States in November, but in Canada on the second Monday in October; elsewhere around the world it may not be a holiday at all. Calling February a winter month does not make sense to someone in New Zealand, for whom it is a summer one.

9. Keep your sentences simple and easy to understand. Short, direct sentences will cause a reader whose native language is not English the least amount of trouble. A good rule of thumb is that the shorter and less complicated your sentences, the easier they will be for a reader to process. Long (more than fifteen words) and complex (multiclause) sentences can be so difficult for readers to unravel that they may skip over them or guess at your message. Do not, however, be insultingly childish as if you were writing for someone in kindergarten. Always try to avoid the passive voice; it is one of the most difficult sentence patterns for a non-native speaker to comprehend. Stick to the common subject-verb-object pattern as often as possible (see Chapter 2, p. 39).

Respecting Readers' Nationality and Ethnic/Racial Heritage

Do not risk offending any of your readers, whether they are native speakers of English or not, with language that demeans or stereotypes their nationality or ethnic and racial background. Here are some precautions to take.

1. Respect your reader's nationality. Always spell your reader's name and country properly, which may mean adding diacritical marks (e.g., accent marks) not used in English. If your reader has a hyphenated last name (e.g., Arana-Sanchez), it would be rude to address him or her by only part of the name (e.g., only Arana or only Sanchez). In addition, be careful not to use the former name of your reader's country or city, for instance, the Soviet Union (now Russia), Malaya (now Malaysia), Czechoslovakia (now the Czech Republic), Rhodesia (now Zimbabwe) or Bombay (now Mumbai). Not only is it rude, but it also demonstrates a lack of care about researching your reader's nationality.

2. Observe your reader's cultural traditions. Cultures differ widely in the way they send and receive information and how they prefer to be addressed, greeted, and informed in a letter. What is acceptable in one culture may be offensive in another. A sales letter to an Asian business executive, for example, needs to employ a very different strategy from one intended for an American reader. The best strategy for an American audience would be hard-hitting and to the point, stressing your product's strengths versus the competitor's weaknesses. But the East Asian way of writing such a letter would be more subtle, indirect, and complimentary.

American:	Our Imaging 500 delivers much more extensive internal imaging than any of our competitors' equipment.
East Asian:	One of the ways we may be able to serve you is by informing you about our new Imaging 500 MRI (Magnetic Resonance Imaging) equipment.

The hard sell in the American example would be a sign of arrogance, suggesting inequality, for a reader in China, Malaysia, or Korea who is more comfortable with a compliment or a wish for prosperity. If given a Korean executive's business card, do not fold or write on it or shove it in your pocket.

3. Honor your reader's place in the world economy. Phrases such as "third-world country," "emerging nation," and "undeveloped/underprivileged area" are derogatory. Using such phrases signals that you regard your reader's country as inferior. Use the name of your reader's country instead. Saying that someone lives in the Far East implies that the United States, Canada, or Europe is the center of culture, the hub of the business community. Never use the word "Oriental," which is insulting. Simply say "East Asia."

4. Avoid insulting stereotypes. Expressions such as "oil-rich Arabs," "time-relaxed Latinos," and "aggressive foreigners" unfairly characterize particular groups. Similarly, prune from your communications any stereotypical phrase that insults one group or singles it out for praise at the expense of another—"Mexican standoff," "Russian roulette," "Chinaman's chance," "Irish wake," "Dutch treat," "Indian giver." The word *Indian* refers to someone from India; use *Native American* to refer to the indigenous people of North America, who want to be known by their tribal affiliations (e.g., the Cherokee).

5. Be visually sensitive to the cultural significance of colors. Do not offend your audience by using colors in a context that would be offensive. Green and orange have a strong political context in Ireland. In China, white does not symbolize purity and weddings but mourning and funerals. Similarly, in India if a married woman wears all white, she is inviting widowhood. A signature or a note written in red would signify anger to an Indonesian reader.

6. Be careful, too, about the symbols you use for international readers. Triangles are associated with anything negative in Hong Kong, Korea, and Taiwan. Political symbols, too, may have controversial implications (e.g., the hammer and sickle, a crescent). Avoid using the flag of a country as part of your logo or letterhead for global audiences. Many countries see this as a sign of disrespect, especially Saudi Arabia, whose flag features the name of Allah.

Writing to Readers from a Different Culture: Some Examples

Figures 4.20 and 4.21 illustrate the correct and incorrect way to write to your international readers. Note how the letter shown in Figure 4.20 violates the guidelines for writing to an international reader:

- uses incorrect format in date line
- misspells name of reader's city
- excludes important postal information
- omits part of reader's surname

The letter is filled with U.S. idioms—*drop you a line, give you a ring*—and uses troubling abbreviations, for instance, *prod. eff. quotas.* Moreover, the writer disregards how the reader records time and temperature—1:30 for an Argentinian reader would signal 1:30 a.m. according to the twenty-four-hour clock; and the low 80's would mark a scorcher on the Celsius scale used in Buenos Aires. Equally disrespectful, the writer's overall tone is condescending ("south of the border") and, at the end of the second paragraph, tells the reader that the U.S. company is superior to the Argentinean one.

See how the letter in Figure 4.21 respects the reader's culture:

- correctly spells and punctuates address
- uses clear date line
- includes appropriate salutation and close
- is clear, courteous, and acknowledges reader's position of authority in company

Above all, the writer respects his audience's culture and role in the business world, and seeks to win the reader's confidence and cooperation, two invaluable assets in the global marketplace.

Figure 4.20　An inappropriately written letter for an international reader.

Pro-Tech, Ltd. 452 West Main St.　Concord, MA 01742　978-634-2756
www.protech.com

Misleading date line

5-8-09

Incorrect, misspelled address

Mr. Antonio Guzman
Canderas
Mercedes Ave.
Bunos Aires, ARG.

Too informal

Dear Tony,

I wanted to drop you a line before the merger hits and in doing so touch base and give you the lowdown on how our department works here in the good old U.S. of A.

Culturally condescending, disrespectful

None of us had a clue that Pro-Tech was going to go south of the border but your recent meeting about the Smartboard T-C spoke volumes to the tech people who praised your operations to the hilt. So it looks like you and I both will be getting a new corp. name. Olé. I love moving from Pro-Tech, Ltd. to Pro-Tech International. We are so glad we can help you guys out.

Filled with American idioms, abbreviations

At any rate, I'm sending you an e-mail with all the ins and outs of our department struc., layout, employees, and prod. eff. quotas. From this info, I'm hoping you'll be able to see ways for us to streamline, cooperate, and soar in the market. I understand that all of this is in the works that you and I need to have a face-to-face and so I'd appreciate your reciprocating with all the relevant data stat.

Disregards time differences and reader's 24-hour clock

Consequently, I guess I'll be flying down your way next month. Before I take off, I would like to give you a ring. How does after lunch next Thursday (say, 1:00–1:30) sound to you? I hope this is doable.

Ignores Fahrenheit/ Celsius temperature scale differences

We've had a spell of great weather here (can you believe it's in the low 80's today!). So, I guess I'll just sign off, and wait 'til I hear from you further.

I send you felicitations and want things to go smoothly before the merger is upon us.

Sounds insincere

Adios,

Frank Sims

Frank Sims

An appropriate revision of Figure 4.20.   **Figure 4.21**

Pro-Tech, Ltd. 452 West Main St.   Concord, MA 01742   978-634-2756
www.protech.com

8 May 2009

Clear date line

Señor Antonio Mosca-Guzman
Director, Quality Assurance
Tecnología Canderas, S.A.
Av. Martin 1285, 4° P.C.
C1174AAB BUENOS AIRES
ARGENTINA

Complete and correct address

Dear Señor Mosca-Guzman:

Courteous salutation

As our two companies prepare to merge, I welcome this opportunity to write
to you. I am the manager for the Quality Assurance division at Pro-Tech, a
title I believe you have at Tecnología Canderas. I am looking forward to
working with you both now and after our companies merge in two months.

Clear and diplomatic opening

Allow me to say that we are very honored that your company is joining ours.
Tecnología Canderas has been widely praised for the research and production
of your Smartboard T-C systems. I know we have much to learn from you,
and we hope you will allow us to share our systems analyses with you. That
way everyone in our new company, Pro-Tech International, will benefit from
the merger.

Respectful view of reader's company

Later this week, I will send you a report about how we manage our division.
I will describe how our division is structured and the quality assurance
inspections we make. I will also give you a brief biography of our staff so
that you can learn about their qualifications and responsibilities.

Explanation of business proce-dures in plain English

The director of our new company, Dr. Suzanne Nknuma, asked me to meet
with you before the merger occurs to discuss how we might help each other.
I would very much like to travel to Buenos Aires in the next two months to
visit with you and take a tour of your company.

Uses plain English

Would you please let me know by e-mail when it may be convenient for us
to talk on the telephone so we might discuss the agenda for our meeting?
I am always in my office from 11:00 to 17:00 Buenos Aires time.

Recognition of reader's time zone

I look forward to working with and meeting you.

Respectfully,

Frank Sims

Frank Sims
Quality Assurance Manager

Polite close

Sometimes your international client may be in the United States. Writers need to exercise the same diplomacy when communicating with international audiences within the United States as they would when communicating with audiences in other countries. Patrice St. Jacques, in Figure 4.22, writes an effective sales letter by zeroing in on Etienne Abernathy's ethnic pride and heritage. Although Abernathy's company is located in the United States, St. Jacques persuasively sees Abernathy from a much broader cultural perspective.

Figure 4.22   A sales letter that appeals to a U.S. international audience.

Distinctive, functional letterhead

4700 Cyprus Avenue
Philadelphia, PA 19172

6 July 2009

Mr. Etienne Abernathy, President
Seagrove Enterprises
1800 S. Port Haven Road
Philadelphia, PA 19103-1800

Dear Mr. Abernathy:

Compliments reader on award

Congratulations on winning the Hanover Award for Community Service. We in the Port Haven area are proud that a business with Caribbean roots has received such a distinguished honor.

Appeals to reader's ethnic background in art, music, and food

To celebrate your and Seagrove's success, as well as all your business entertaining needs (annual banquet, monthly meetings, etc.), I invite you to Island Jacques. We are a family-owned business that for 30 years has offered Philadelphia residents the finest Caribbean atmosphere and food west of the Islands. Our black pepper shrimp, reggae or mango chicken, and steak St. Lucie—plus our irresistible beef and pork jerk—are the talk from here to Kingston. You and your guests can also savor our original Caribbean art and enjoy our steel drum music.

Features flexible dining options, including a specialized ethnic menu

Island Jacques can offer Seagrove a variety of dining options; with separate rooms, we are small enough for an intimate party of 4 yet large enough to accommodate a group of 250. We can do early lunches or late dinners, depending on your schedule. And we even cater, if that's your style. Our chefs—Diana Maurier and Emile Danticat—will prepare a special calypso menu just for you. Also a benefit, our prices are generously competitive for the Philadelphia area.

www.islandjacques.netdoor.com
856-555-3295

Continued

Page 2

Please call me soon so you can see Island Jacques's unique hospitality.
For your convenience, I am enclosing a copy of this week's menu
delights. Check out our website, too, for a taste of Caribbean sound. We
would love to feature Seagrove as Island Jacques's "Guest of the Week"!

Stay Cool, Mon

Patrice St Jacques

Patrice St. Jacques
Manager

*Gives reader
incentive, tied
to his heritage,
to act soon*

Writing letters with the specific needs of your audience in mind is a crucial skill to have in the
world of work.

✔ Revision Checklist

- ☐ Planned what I am going to say to my readers. Did necessary homework and
double-checking to answer any questions. Proved to my readers that I am
knowledgeable about my topic.
- ☐ Used an appropriate (and consistent) format for my letters.
- ☐ Included all parts of a letter.
- ☐ Printed letter with generous margins to make it neat and professional, not cramped
or lopsided.
- ☐ Signed and proofread my letter.
- ☐ Organized information in letters in the most effective way for my message and
for my readers.
- ☐ Emphasized the "you attitude" with my readers, whether employer,
customer/client, or co-worker.
- ☐ Conveyed impression of being courteous, professional, and easy to work with.
- ☐ Used clear and concise language appropriate for my reader.
- ☐ Cut out anything that sounded flowery, stuffy, or bureaucratic.
- ☐ Began my correspondence with reader-effective strategies. If reporting good
news, told the reader right away. If reporting bad news, was diplomatically indi-
rect and considerate of my reader's reactions.
- ☐ Followed the four A's of effective sales letters; identified and convinced my tar-
get audience.

Continued

☐ Wrote complaint letters in a calm and courteous tone. Informed the reader what is wrong, why it is wrong, and how the problem should be solved.

☐ Wrote adjustment letters that say "Yes" sincerely and to the point. Made those that say "No" fair. Acknowledged reader's point of view and provided clear explanation for my refusal.

☐ Sent appropriate collection letters based on audience and time overdue.

☐ Did appropriate research about reader's culture, in print sources, through online sources, and with multinational colleagues or through foreign language instructors.

☐ Adopted a respectful, not condescending, tone.

☐ Avoided cultural taboos considered offensive by my reader.

☐ Chose colors and symbols culturally appropriate for my reader and the context of my message.

☐ Used plain and clear language that my reader would understand.

☐ Selected the right format, salutation, and complimentary close for my reader.

☐ Observed the reader's units of measurement for time, temperature, currency, dates, and numbers.

Exercises

Additional Activities related to writing letters are located at **college .cengage .com/pic/ kolinconcise2e.**

1. Write appropriate inside addresses and salutations to (a) a woman who has not specified her marital status, (b) an officer in the armed forces, (c) a professor at your school, (d) an assistant manager at your local bank, (e) a member of the clergy, (f) your congressional representative.

2. Rewrite the following sentences to make them more personal.
   a. It becomes incumbent on this office to cancel order #2394.
   b. Management has suggested the curtailment of parking privileges.
   c. ALL USERS OF HYDROPLEX: Desist from ordering replacement valves during the period of Dec. 20–30.
   d. The request for a new catalog has been honored; it will be shipped to same address soon.
   e. Perseverance and attention to detail have made this writer important to company in-house work.
   f. The Director of Nurses hereby notifies staff that a general meeting will be held Monday afternoon at 3:00 p.m. sharp. Attendance is mandatory.
   g. Reports will be filed by appropriate personnel no later than the scheduled plans allow.

3. The following sentences from letters are discourteous, boastful, vague, or lacking the "you attitude." Rewrite them to correct those mistakes.
   a. Something is obviously wrong in your head office. They have once more sent me the wrong model number. Can they ever get things straight?

b. My instructor wants me to do a term paper on safety regulations at a small factory. Since you are the manager of a small factory, send me all the information I need at once. My grade depends heavily on all this.

c. It is apparent that you are in business to rip off the public.

d. I have waited for my confirmation for two weeks now. Do you expect me to wait forever or can I get some action?

e. It goes without saying that we cannot honor your request.

f. Your application has been received and will be kept on file for six months. If we are interested in you, we will notify you. If you do not hear from us, please do not write us again. The soaring costs of correspondence and the large number of applicants make the burden of answering pointless letters extremely heavy.

g. My past performance as a medical technologist has left nothing to be desired.

h. Credit means a lot to some people. But obviously you do not care about yours. If you did, you would have sent us the $249.95 you rightfully owe us three months ago. What's wrong with you?

4. Write a business letter to one of the following individuals.
   a. your mayor, asking for an appointment and explaining why you need one
   b. your college president, stressing the need for more parking spaces or for additional computer terminals in a library
   c. the local water department, asking for information about fluoride supplements
   d. an editor of a weekly magazine, asking permission to reprint an article in a company newspaper
   e. a disc jockey at a local radio station, asking for more songs by a certain group
   f. a computer vendor, asking about costs and availability of software packages; explain your company's special needs

5. Write a letter of inquiry to a utility company, a safety or health care agency, or a company in your town and ask for a brochure describing its services to the community. Be specific about your reasons for requesting the information.

6. Choose one of the following and write a sales letter addressed to an appropriate audience on why they should
   a. work for the same company you do
   b. live in your neighborhood
   c. be happy taking a vacation where you did last year
   d. dine at a particular restaurant
   e. use a particular software program
   f. have their cars repaired at a specific garage
   g. give their real estate business to a particular agency
   h. visit your Web site

7. Rewrite the following sales letter to make it more effective. Add any details you think are relevant.

```
Dear Pizza Lovers:

Allow me to introduce myself. My name is Rudy Moore and I am the
new manager of Tasty Pizza Parlor in town. The Parlor is located
```

at the intersection of North Miller Parkway and 95th Street. We are open from 10 a.m. to 11 p.m., except on the weekends, when we are open later.

I think you will be as happy as I am to learn that Tasty's will now offer free delivery to an extended service area. As a result, you can get your Tasty Pizza hot when you want it.

Please see your weekly newspapers for our ad. We also are offering customers a coupon. It is a real deal for you.

I know you will enjoy Tasty's pizza and I hope to see you. I am always interested in hearing from you about our service and our fine product. We want to take your order soon. Please come in.

8. Send a follow-up letter to one of the following individuals:
   a. a customer who informs you that she will no longer do business with your firm because your prices are too high
   b. a family of four who stayed at your motel for two weeks last summer
   c. a wedding party that used your catering services last month
   d. a customer who exchanged a coat for the purchase price
   e. a customer who purchased a used car from you and who has not been happy with warranty service
   f. a company that bought software from you nine months ago, alerting them to improvements in the software

9. Write a complaint letter about one of the following:
   a. an error in your utility, telephone, or credit card bill
   b. discourteous service you received on an airplane or bus
   c. a frozen food product of poor quality
   d. a shipment that arrived late and damaged
   e. an insurance payment to you that is $250.00 less than it should be
   f. a television station's policy of not showing a particular series
   g. junk mail or spam that you are receiving
   h. equipment that arrives with missing parts
   i. misleading representation by a salesperson
   j. incorrect information given at a Web site

10. The following story appeared recently in a local newspaper. Based on information in this story, which you may want to supplement, write the following complaint/adjustment letters:
    a. a complaint letter to the city from resident Jo Souers
    b. a complaint letter to Finicky Pet Food from city officials warning about dangers of odors to the residential area
    c. a letter from plant manager Dean Niemann to the residents of Bienville Place subdivision
    d. a letter from city officials to the residents of Bienville Place subdivision

Residents concerned about relocation of pet food plant

OCEAN SPRINGS (AP) – Finicky Pet Food is moving its processing plant from Pascagoula to Ocean Springs, a decision that has some residents concerned of possible odor and other problems.

The plant is moving to an industrial area bordering a subdivision of expensive homes.

"The wind doesn't discriminate," said Jo Souers, who lives in the Bienville Place subdivision. "I don't want this in our neighborhood."

City officials said the plant is moving to an area zoned to accommodate it.

"We don't have a lot of control over it," said city planner Donovan Scruggs. "It is a permitted use for this property."

Scruggs said the property was zoned industrial before the subdivision was built. A body shop, cabinet shop, and boat business are located nearby.

The plant will be built in the small industrial area on U.S. 90, directly across the highway from the Super Wal-Mart.

It is moving into a vacant building, the interior of which has been renovated for its new purpose, city officials said.

The plant will process frozen fish and fish parts for bait and pet food. It will employ 10 workers, with that number doubling during fishing season.

Plant manager Dean Niemann said in a statement that the company no longer needed its Pascagoula location near deep water, which was rented from the county.

Scruggs said the city has investigated the possibility that the plant will emit odors.

"We've told Dean (Niemann) from day one, 'You're locating next to a residential area. If you start stinking, action will be taken,'" Scruggs said.

He said the city has a nuisance ordinance that should handle anything that might arise.

Reprinted with permission of The Associated Press.

---

11. Rewrite the following ineffective adjustment letter saying "No."

Dear Customer:

Our company is unwilling to give you a new toaster or to refund your purchase price. After examining the toaster you sent to us, we found that the fault was not ours, as you insist, but yours.

Let me explain. Our toaster is made to take a lot of punishment. But being dropped on the floor or poked inside with a knife, as you probably did, exceeds all decent treatment. You must be careful if you expect your appliances to last. Your negligence in this case is so bad that the toaster could not be repaired.

In the future, consider using your appliances according to the guidelines set down in warranty books. That's why they are written.

```
Since you are now in the market for a new toaster, let me sug-
gest that you purchase our new heavy-duty model, number 67342,
called the Counter-Whiz. I am taking the liberty of sending
you some information about this model. I do hope you at least
go to see one at your local appliance center.

Sincerely,
```

12. You are the manager of a computer software company, and one of your sales-people has just sold a large order to a new customer whose business you have tried to obtain for years. Unfortunately, the salesperson made a mistake writing out the invoice, undercharging the customer by $229. At that price, your company would not break even, so you must write a letter explaining the problem so that the customer will not assume all future business dealings with your firm will be offered at such "below market" rates. Decide whether you should ask for the $229 or just "write it off" in the interest of keeping a valuable new customer.
    a. Write a letter to the new customer, asking for the $229 and explaining the problem while still projecting an image of your company as accurate, professional, and very competitive.
    b. Write a letter to the new customer, not asking for the $229 but explaining the mistake and emphasizing that your company is both competitive and professional.
    c. Write a letter to your boss explaining why you wrote letter **a.**
    d. Write a letter to your boss explaining why you wrote letter **b.**
    e. Write a letter to the salesperson who made the mistake, asking him or her to take appropriate action with regard to the new customer.

13. Write a sales letter similar to the one in Figure 4.22 from a manager of one of the following ethnic restaurants or one of your choice. Make sure you include relevant details for the particular audience:
    a. Mexican   j. German
    b. Indian   k. Irish
    c. Cuban   l. Chinese
    d. Soul food   m. Pakistani
    e. Czech   n. Italian
    f. Turkish   o. Argentine
    g. Vietnamese   p. French
    h. Thai   q. Afghan
    i. Greek   r. Hindi

14. Rewrite the following letters, making them appropriate for a reader whose native language is not English. As you revise the letters, pay attention to the words, measurements, and sentence constructions you employ. Be sure to consider the reader's cultural traditions by omitting any cultural insensitivity.

    a. Dear Pal,

```
Our stateside boss hit the ceiling earlier today when she
learned that our sales quota for this quarter fell precipi-
```

tously short. Ouch! Were I in her spot, I would have exploded too. Numerous missives to her underlings warned them to get off the dime and on the stick, but they were oblivious to such. These are the breaks in our business, right? We can't all bat 1000.

Let's hope that next quarter's sales take a turn for the best by 12-1-08. If they are as disastrous, we all may be in hot water. Until then, we will have to watch our p's and q's around here. We're freezing here—20° today.

Cheers,

b. Dear Margaret Wong,

It's not every day that you have the chance to get in on the ground floor of a deal so good you can actually taste it. But Off-Wall Street Mutual can make the difference in your financial future. Give me a moment to convince you.

By becoming a member of our international investing group for just under $250, you can just about ensure your success. We know all the ins and outs of long-term investing and can save you a bundle. Our analysts are the hot shots of the business and always look long and hard for the most propitious business deals. The stocks we select with your interests in mind are as safe as a bank and not nearly so costly for you. Unlike any of your undertrained local agents, we can save you money by investing your money. We are penny pinchers with our clients' initial investments, but we are King Midas when it comes to transforming those investments into pure gold.

I am enclosing a brochure for you to study, and I really hope you will examine it carefully. You would be foolish to let a deal like Off-Wall Street Mutual pass you by. Go for it. Call me by 3:00 today.

Hurriedly,

15. Interview a student at your school or a co-worker who was born and raised in a non-English-speaking country about the proper etiquette in writing a business letter to someone from his or her country. Collaborate with that student to write a letter (for example, a sales letter or a letter asking for information) to an executive from that country.

16. In a letter to your instructor, describe the kinds of adaptations you had to make for the international reader you wrote to in Exercise 15.

How to Get a Job: Résumés, Letters of Application, and Interviews

To expand your understanding of how to get a job, take advantage of the Web Links, Additional Activities, and ACE Self-Tests at **college.cengage .com/pic/ kolinconcise2e**.

Obtaining a job today involves a lot of hard work. Before your name is added to a company's payroll, you will have to do more than simply walk into the human resources office and fill out an application form. Finding the *right* job takes time. And finding the right person to fill that job also takes time for the employer.

Steps the Employer Takes to Hire

From the employer's viewpoint, the stages in the search for a valuable employee include the following:

1. deciding what duties and responsibilities go with the job and determining the qualifications the future employee should possess
2. advertising the job on the company website, in newspapers, and in professional publications
3. reading and evaluating online and hard-copy résumés and letters of application
4. having candidates complete application forms
5. requesting further proof of candidates' skills (letters of recommendation, transcripts, a portfolio of graphics, reports, and other documents)
6. interviewing selected candidates
7. offering the job to the best-qualified individual

Sometimes the steps are interchangeable, especially steps 4 and 5, but generally speaking, employers go through a long and detailed process to select employees. Step 3, for example, is among the most important for employers (and the most crucial for job candidates). At that stage employers often classify job seekers into one of three groups: those they definitely want to interview, those they may want to interview, and those in whom they have no interest.

Steps to Follow to Get Hired

As a job seeker you will have to know how and when to give the employer the kinds of information the preceding seven steps require. You will also have to follow a schedule that has major deadlines in your search for a job. The following six procedures will be required of you:

1. analyzing your strengths and restricting your job search
2. looking in the right places for a job
3. preparing a résumé
4. writing a letter of application
5. filling out a job application
6. going to an interview

Your timetable should match that of your prospective employer. This chapter shows you how to begin your job search and how to prepare an appropriate résumé and letter that are a part of your job search.

Analyzing Your Strengths

Before you apply for jobs, analyze your job skills, career goals, and interests. Here are some points to consider.

1. Make an inventory of your most significant accomplishments in your major and/or on the job. What are your greatest strengths—writing and speaking, working with people in small groups, organizing and problem solving, speaking a second language, managing money, troubleshooting, creating graphics?
2. Decide which specialty within your chosen career appeals to you the most. If you are in a nursing program, do you want to work in a large teaching hospital, for a home health or hospice agency, or in a physician's office? What kinds of patients do you prefer to care for—pediatric, geriatric, psychiatric?
3. What are the most rewarding prospects of a job in your profession? What most interests you about a position—travel, technology, international contacts, on-the-job training, helping people, being creative?
4. What are some of the greatest challenges you face in your career today—or will in five years?
5. Which specific companies or organizations have the best track record in hiring and promoting individuals in your field? What qualifications will such firms insist on from prospective employees?

The *Occupational Outlook Handbook* (*http://www.bls.gov/oco/*) can give you valuable career information on job prospects, requirements, salary ranges, and contacts.

Once you answer the previous questions you can avoid applying for positions for which you are either overqualified or underqualified. If a position requires ten years of related work experience and you are just starting out, you will only waste the employer's time and your own by applying. However, if a job requires a certificate or license and you are in the process of obtaining one, go ahead and apply.

▌ Looking in the Right Places for a Job

One way to search for a job is simply to send out a batch of letters or online résumés to companies you want to work for. But how do you know what jobs, if any, those companies have available, what qualifications they are looking for, and what deadlines they might want you to follow? You can avoid these uncertainties by (a) knowing where to look for a job, and (b) understanding knowing what a specific job entails. Consult the following resources for a wealth of job-related information.

1. **Networking.** Networking pays. It is regarded as the most important strategy to follow. John D. Erdlen and Donald H. Sweet, experts on job searching, cite the following as a primary rule of job hunting: "Don't do anything yourself you can get someone with influence to do for you." Let your professors, friends, classmates, neighbors, relatives, and even your clergy know you are looking for a job. They may hear of something and can notify you or, better yet, recommend you for the position. See how the job seekers in Figures 5.8 (p. 155) and 5.9 (p. 157) have successfully networked with people who can assist them in their job search. You can also network with people you don't know personally through websites such as:

http://www.linkedin.com

http://www.ryze.com

http://www.tribe.net

Also, attend job fairs, professional and organization meetings, community and civic functions—to meet the right contact people whom you can ask for advice and also for possible follow-up help and recommendations.

2. **The Internet.** Learn about jobs by visiting a company's website to see if it has vacancies and what the qualifications are for them. You can also consult the many online job services that list positions and sometimes give advice, including:

- Monster.com—*http://www.monster.com*
- Riley Guide—*http://www.rileyguide.com*
- Yahoo! HotJobs—*http://www.hotjobs.yahoo.com*
- Career Builder—*http://www.careerbuilder.com*
- Job.com—*http://www.job.com*
- Career.com—*http://www.career.com*
- Job Central.com—*http://www.jobcentral.com*

There are also many specialized job search sites. For a helpful list broken down by job area, visit the University of Delaware Career Services Center's "Specialized Job Sites" page at *http://www.udel.edu/CSC/specnet.html*.

Make sure you always sign up for job alerts. Also study ways to use search engines to find contacts and jobs.

3. Newspapers. Look at local newspapers as well as the Sunday editions of large city papers with a wide circulation, such as the *New York Times Job Market* (*http://www.jobs.nytimes.com*). The *National Business Employment Weekly* (*http://www.careerjournal.com*), published by the *Wall Street Journal,* also lists jobs in different areas, including technical and managerial positions.

4. Your campus placement office. Counselors keep an online file of available positions and can also notify you when a firm's recruiter will be on campus to conduct interviews. Many placement offices have recruiting databases, allowing students access to a broad range of recruiting contacts and interview information. They can also help you locate summer and part-time work, both on and off campus, positions that might lead to full-time jobs. Most important, they will give you sound advice on your job search, including strategies for finding the right job, salary ranges, and interview tips. Many placement offices also sponsor career fairs to bring job seekers and employers together in specific professional, technical fields.

5. Federal and state employment offices. The U.S. government is one of the biggest employers in the country. During 2007 and 2008, for instance, the most active career site on the Internet was operated by the federal government, with 1.7 million new hires. Counselors at federal and state employment centers also help job seekers find career opportunities. Figure 5.1 (p. 136) shows the home page of the website for USAJOBS, which helps job seekers find employment opportunities with the U.S. government. Consult the following websites for listings of government jobs:

- USA Jobs—*http://www.usajobs.gov*
- Federal Jobs.Net—*http://www.federaljobs.net*
- Student Jobs—*http://www.studentjobs.gov*

6. Professional and trade journals plus associations in your major. Identify the most respected periodicals in your field and search their ads. Each issue of *Food Technology* features a section called "Professional Placement," a listing of jobs all over the country. Similarly, *CIO Magazine—IT Professional Research Center* (*http://www.cio.com/research/itcareer/*) can help you find jobs in computer engineering and technology. The *Encyclopedia of Associations* (*http://library.dialog.com/bluesheets/html/bl0114.html*) also lists journals in your profession.

7. The human resources department of a company or agency you would like to work for. Often you will be able to fill out an application even if there is not a current opening. But, do not call employers asking about openings; a visit shows a more serious interest.

8. A résumé database service. A number of online services will put your résumé in a database and make it available to prospective employers, who scan the database regularly to find suitable job candidates. Figure 5.2 describes a résumé database service offered by one professional organization—the Association for Computing

Figure 5.1   USAJOBS website.

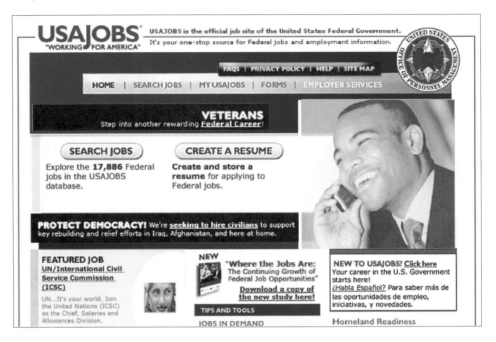

Machinery (ACM)—for its members. Check to see if a professional society to which you belong (or might join) offers a similar service.

9. Professional employment agencies. Some agencies list jobs you can apply for free of charge (because the employer pays the fee) while others charge a stiff fee, usually a percentage of your first year's salary. Be sure to ask who pays the fee for their service. Because employment agencies often find out about jobs through channels already available to you, speak to your campus career center first.

▌ Preparing a Résumé

The résumé, sometimes called a **curriculum vitae,** may be the most important document you prepare for your job search. It deserves your utmost attention. A résumé is not your life history or your emotional autobiography, nor is it a transcript of your college work. It is a factual and concise summary of your qualifications, showing a prospective employer that you have what it takes (in education and experience) to do the job you are applying for. Regard your résumé as a persuasive ad for your professional qualifications.

What you include—your key details, the wording, the ordering of information, and formatting—are all vital to your campaign to sell yourself and land the inter-

---

Employers Benefit from ACM Résumé Database

The ACM Résumé Database is composed of résumés of ACM members—high-caliber, information-technology professionals who can bring expertise to your company. All résumés are up-to-date and can be searched according to criteria you provide, in a short period of time and at low rates. Single searches as well as annual subscriptions to the database can be requested from the database administrator, Resume-Link, at 614-529-0429. ACM institutional members get a 10% discount off the cost of the search.

And for ACM members! Our database also is now searchable for internships and co-op positions, as well as for full-time and consultant placements. Use this free career development service and submit your résumé on-line at **http://www.Resume-Link.com/**.

---

view. Employers want to see the most crucial and current details about your qualifications quickly. Accordingly, keep your print résumé short (preferably one page, never longer than two) and hard hitting. The same thing goes for your online résumé. Everything on your résumé needs to convince an employer you have the exact skills and background he or she is looking for. See page 149 for an online résumé.

What Employers Like to See in a Résumé

Prospective employers will judge you and your work by your résumé, their first view of you and your qualifications. They want to see the following seven characteristics in an applicant's résumé.

- **Honesty.** Be truthful about your qualifications—your education, experience, and skills. Distorting, exaggerating, or falsifying information about yourself in your résumé is unethical and could cost you the job. If you were a clerical assistant to an attorney, don't describe yourself as a paralegal. Employers demand trustworthiness.
- **Attractive.** The document should be pleasing to the eye with appropriate spacing, typeface, and use of boldface; it shows you have a sense of proportion and document design and that you are visually smart.
- **Carefully organized.** The orderly arrangement of information is easy-to-follow, logical, and consistent; it shows you have the ability to process information and to summarize. Employers prize analytical thinking.

- **Concise.** Generally, keep your résumé to one page, as in Figure 5.3 (p. 140). However, depending on your education or job experience, you may want to include a second page. Résumés are written in short sentences that omit "I" and that use action-packed verbs, such as those listed in Table 5.1.
- **Accurate.** Grammar, spelling, dates, names, titles, and programs are correct; typos, inconsistencies, and math errors say you don't check your facts and figures.
- **Current information.** All information is up-to-date and documented, with no gaps or sketchy areas. Missing or incorrect dates or not supplying contact information are flags to employers to reject your résumé.
- **Relevance.** The information is appropriate for the job level, shows that you have the necessary education and experience, and confirms that you can be an effective team player.

Your goal is to prepare a résumé that shows the employer you possess the sought-after job skills. One that is unattractive, difficult to follow, poorly written, filled with typos and spelling mistakes or that are not relevant for the prospective employer's needs just will not make the first cut.

It might be to your advantage to prepare several versions of your résumé and then adapt each one you send out to the specific job skills a prospective employer is looking for. It pays to customize your résumé. Following the process in the next section will help you prepare any résumé.

TABLE 5.1   Action Verbs to Use in Your Résumé

accommodated	created	installed	reported
accomplished	customized	instituted	researched
achieved	dealt in	instructed	saved
adapted	designed	interned	scheduled
adjusted	determined	interpreted	searched
administered	developed	maintained	selected
analyzed	directed	managed	served
arranged	drafted	mapped	settled
assembled	earned	monitored	sold
assisted	elected	motivated	solved
attended	established	navigated	supervised
awarded	estimated	negotiated	taught
budgeted	evaluated	operated	tested
built	expanded	organized	tracked
calculated	expedited	oversaw	trained
coached	figured	performed	translated
collected	fulfilled	planned	tutored
communicated	guided	prepared	updated
compiled	handled	programmed	upgraded
completed	headed	provided	verified
composed	implemented	reappraised	weighed
computed	improved	received	wired
conducted	increased	reconciled	won
converted	informed	reduced	worked
coordinated	initiated	re-evaluated	wrote

The Process of Writing Your Résumé

To write an effective résumé, ask the following important questions:

1. What classes did you excel in?
2. What papers or reports earned you your highest grades?
3. What computer skills have you mastered—languages, software knowledge, navigating groupware, and developing Net resources? Knowledge of e-commerce? Ability to design a blog and website?
4. What jobs have you had? For how long and where? What were your primary duties?
5. What technical skills have you acquired?
6. Do you work well with people? What skills do you possess as a member of a team working toward a common job goal (e.g., finishing a report).
7. Can you organize complicated tasks or solve problems quickly?
8. Have you had experience/responsibilities managing money—collecting dues for a student or civic group, preparing payrolls, conducting audits?
9. Have you won any awards or scholarships or received a raise, bonus, commendation, and/or promotion at work or in the military?

Pay *special attention* to your four or five most significant, job-worthy strengths and work especially hard on listing them concisely.

Although not everything you have done relates directly to a particular job, indicate how your achievements are relevant to the employer's overall needs. For example, supervising staff in a convenient store points to your ability to perform the same duties in another business context.

Balancing Education and Experience

If you have years of experience, don't flood your prospective employer with too many details. You cannot possibly include every detail of your job(s) for the last ten or twenty years.

- Emphasize only those skills and positions most likely to earn you the job.
- Eliminate your earliest jobs that do not relate to your present employment search.
- Combine and condense skills acquired over many years and through many jobs.

Figure 5.5 (p. 145) shows the résumé of Dora Cooper Bolger who has a great deal of community experience to offer prospective employers.

Many job candidates who have spent most of their lives in school are faced with the other extreme: not having much job experience to list. The worst thing to do is to write "None" for experience. Any part-time internship, summer, or other seasonal jobs, as well as work done for a library or science laboratory, can convince an employer that you are responsible and knowledgeable about the obligations of being an employee. Figure 5.3 shows a résumé from Anthony Jones, a student with very little job experience; Figure 5.4 (pp. 141–142) shows the one of María Lopez, a student with a few years of experience.

Figure 5.3   Résumé from a student with little job experience.

Anthony H. Jones
73 Allenwood Boulevard
Santa Rosa, California 95401-1074
707-555-6390
ajones@plat.com
www.plat.com/users/ajones/resume.html

WEBSITE DEVELOPER
DESIGNER
GRAPHIC ARTIST

Offers precise, convincing objective

Career Objective

Full-time position as a layout artist with a commercial publishing house using my knowledge of state-of-the-art design technology.

Starts with most important qualification— education

Education

Santa Rosa Junior College, 2006–2008, A.S. degree to be awarded in 2008
Dean's List in 2005; GPA 3.45
Major: Commercial Graphics Illustration, with specialty in design layout
Related courses included:
- Digital Photography
- Graphics Programs: Illustrator, Photoshop
- Desktop Publishing: QuarkXPress, WordPerfect Suite 8

Stresses job-related activities of apprenticeship

Internship, 2007–2008, McAdam Publishers
Major projects included:
- Assisting layout editors with page composition and importing images.
- Writing detailed reports on digital photography, designs, and artwork used in *Living in Sonoma County* (www.sonomacounty.com) and *Real Estate in Sonoma County* (www.resc.net) magazines.

Experience

Salesperson (part-time), **2003–2005**, Buchman's Department Store
Duties included assisting customers in sporting goods and appliance departments and coordinating sport shop by displaying merchandise.

Includes part-time work experience

Computer Skills

Know QuarkXPress Passport, WordPerfect X3.

Demonstrates skills in web design

Related Activities

Volunteer; designed website and three-fold brochure for the Santa Rosa Humane Society's 2007 fund drive; helped raise $4,300.

Credentials and portfolio emphasize accomplishments

References

References, college transcripts, and a portfolio of web designs and photographs available upon request.

Résumé from a student with some job experience.   **Figure 5.4**

MARÍA H. LOPEZ
1725 Brooke Street
Miami, Florida 32701-2121
(305) 555-3429   **mlopez@eagle.com**

Career Objective	Position assisting dentist in providing dental care, counseling, and preventive dental treatments, especially in pedodontics.	*Provides easy-to-find contact information*
Education	A.S. in Dental Hygiene, Miami-Dade Community College, Miami, Florida August 2007–May 2009 Completed courses in oral pathology, dental materials and specialties, periodontics, and community dental health. Experienced with procedures and instruments used with oral prophylaxis techniques. Subject of major project was proper nutrition and dental health for preschoolers.	*Lists course and clinical work required for licensure and job*
	Minor: Psychology (twelve hours in child and adolescent psychology) Received excellent evaluations in business writing course. GPA is 3.3. Bilingual: Spanish/English.	*Calls attention to counseling and bilingual skills that benefit employer and patients*
	Will take American Dental Assistants' Examination on June 2.	*Emphasizes taking licensure exam*
Experience	St. Francis Hospital, Miami Beach, Florida April 2005–July 2007 Full-time clerk on the pediatric floor. Ordered supplies, maintained records, transcribed orders, and greeted and assisted visitors.	
	Murphy Construction Company, Miami, Florida June 2004–April 2005 Office assistant-receptionist. Did keyboarding, billing, and mailing in small office (four employees).	*Highlights previous job responsibilities in health care setting*
	City of Hialeah, Florida Summers 2002–2003 Water meter reader	
Computer Skills	Microsoft Office, Excel, Health Care spreadsheet	

Continued

Figure 5.4   (Continued)

Lopez 2

References	The following individuals have written letters of recommendation for my placement file, available from the Placement Center, Miami-Dade Community College, Medical Center Campus, Miami, FL 33127-2225.

Wisely lists professionals well known to prospective/ area employers

Sister Mary Pela
Pediatric Unit
St. Francis Hospital
10003 Collins Avenue
Miami Beach, FL 33141
(305) 555-5113

Professor Mitchell Pelbourne
Department of Dental Hygiene
Miami-Dade Community College
Medical Center Campus
Miami, FL 33127
(305) 555-3872

Tia Gutiérrez, D.D.S.
9800 Exchange Avenue
Miami, FL 33167
(305) 555-1039

Mr. Jack Murphy
1203 Francis Street
Miami, FL 33157
(305) 555-6767

What to Exclude from a Résumé

Knowing what to exclude from a résumé is as important as knowing what to include. Since federal employment laws prohibit discrimination on the basis of age, sex, race, national origin, religion, marital status, or disability, do not include such information on your résumé. Here are some other details best left out of your résumé:

- salary demands, expectations, or ranges
- preferences for work schedules, days off, or overtime
- comments about fringe benefits
- salary demands, expectations, or ranges
- travel restrictions
- reasons for leaving your previous job
- your photograph (unless you are applying for a modeling or acting job)
- social security number
- comments about your family, spouse, or children
- height, weight, hair or eye color
- personal information, hobbies, interests (unless relevant to the job you are seeking, in which case put under Related Skills)

Save comments about salary and schedules for your interview. The résumé should be written appropriately to get you that interview.

Parts of a Résumé

Contact Information

At the top of the page center your name (do not use a nickname); address, including your ZIP code; telephone number; and e-mail address. Avoid unprofessional e-mail addresses such as **Toughguy@netfield.com** or **barbiegirl@techscape.com**. Also, include your website and fax number if you have these for an employer to contact you.

Career Objective

One of the first things a prospective employer reads is your career objective statement that specifies the exact type of job you are looking for and in what ways you are qualified to hold it. Create an objective that precisely dovetails with the prospective employer's requirements. Such a statement should be the result of your focused self-evaluation and your evaluation of the job market (see p. 133). It will influence everything else you include. Depending on your background and the types of jobs you are qualified for, you might formulate two or three different career or employment objectives to use with different versions of your résumé as you apply for various positions.

To write an effective career objective statement, ask yourself four basic questions: (1) What kind of job do I want? (2) What kind of job am I qualified for? (3) What capabilities do I possess? (4) What kinds of skills do I want to learn?

Avoid trite or vague goals such as "looking for professional advancement" or "want to join a progressive company." Compare the vague objectives on the left with the more precise ones on the right.

Unfocused	Focused
Job in sales to use my aggressive skills in expanding markets.	Regional sales representative using my proven skills in marketing and communication to develop and expand a customer base.
Full-time position as staff nurse.	Full-time position as staff nurse on cardiac step-down unit to offer excellent primary care nursing and patient/family teaching.

Credentials

The order of the next two categories—**education** and **experience**—can vary. Generally, if you have lots of experience, or related skills, list it first as Dora Cooper Bolger did (see Figure 5.5). However, if you are a recent graduate short on job experience, list education first, as Anthony Jones did (see Figure 5.3). María Lopez (Figure 5.4) also decided to place her education before her job experience because the job she was applying for required the formal training she received at Miami-Dade Community College.

Education

Begin with your most recent education first, then list everything **significant since high school.** Give the name(s) of the school(s); dates attended; and degree, diploma, or certificate earned. Don't overlook military schools or major training programs (EMT, court reporter), institutes, internships, or workshops you have completed.

Remember, however, that a résumé is not a transcript. Simply listing a series of courses will not set you apart from hundreds of other applicants taking similar courses across the country. Avoid vague titles such as Science 203 or Nursing IV. Instead, concentrate on describing the kinds of skills you learned.

> 30 hours in planning and development courses specializing in transportation, land use, and community facilities; 12 hours in field methods of gathering, interpreting, and describing survey data in reports.

> Completed 28 hours in major courses in business marketing, management, and materials in addition to 12 hours in information science, including HTML/web publishing.

List your grade point average (GPA) only if it is 3.0 or above; otherwise, indicate your GPA in just your major or during your last term, again if it is above 3.0.

Experience

Your job history is the key category for many employers. It shows them that you have held jobs before and that you are responsible. Here are some guidelines about listing your experience.

1. Begin with your most recent position and work backward—in reverse chronological order. List the company or agency name, location (city and state), and your title. Do not mention why you left a job.

2. For each job or activity provide short descriptions of your duties and achievements. Rather than saying you were a secretary, indicate that you wrote business letters, used various software programs, designed a company website, prepared schedules for part-time help in an office of twenty-five people, or assisted the manager in preparing accounts.

3. In describing your position(s), emphasize any responsibilities that involved handling money; managing other employees; working with customer accounts, services, and programs; or writing letters and reports. Prospective employers are interested in your leadership abilities, financial shrewdness (especially if you saved your company money), tact in dealing with the public, and communications skills. They will also be favorably impressed by promotions you may have earned.

4. If you have been a full-time parent for ten years or a caregiver for a family member or friend, indicate the management skills you developed while running a household and any community or civic service, as Dora Cooper Bolger does in her résumé in Figure 5.5. She skillfully relates her family and community accomplishments to the specific job she seeks. Her volunteer work translates into an asset for her prospective employer.

Dora Cooper Bolger's résumé organized by skill areas.   **Figure 5.5**

DORA COOPER BOLGER
1215 Lakeview Avenue
Westhampton, MI 46532
Home: 616-555-4772   **Cell:** 616-555-4773   **dbolger@aol.com**

OBJECTIVE	Seek full-time position as public affairs officer in health care, educational, or nonprofit organization

SKILLS

Organizational Communication
- Delivered 18 presentations to civic groups on educational issues
- Recorded minutes and helped formulate agenda as president of large, local PTA for past 6 1/2 years
- Possess excellent computer skills in PeopleSoft and Microsoft Word
- Updated and maintained computerized mailing lists for Teens in Trouble and Foster Parents' Association

Financial
- Spearheaded 3 major fund-raising drives (total of $178,000 collected)
- Prepared and implemented large family budget (3 children, 8 foster children) for 15 years
- Served as financial secretary, Faith Methodist Church, for 4 years

Administrative
- Organized volunteers for American Kidney Fund (last 5 years)
- Established and oversaw neighborhood carpool (17 drivers; more than 60 children) for 7 years
- Coordinated after-school tutoring program for Teens in Trouble; president since 1996

HONORS	"Volunteer of the Year" (2008), Michigan Child Placement Agency
EDUCATION	Metropolitan Community College, A.A., 2005 Mid-Michigan College, B.S., expected 2009; major: public administration; minor: psychology. GPA 3.45
WORK EXPERIENCE	Payroll clerk, 1995–2003 (full- and part-time): Merrymount Plastics; Foley Co.; Westhampton Health Dept.; G & K Electric
REFERENCES	Available on request

Aptly features skill areas before education

Links achievements and awards from volunteer and home-based work skills most important to employer

Chooses strong, active verbs to convey image of a results-oriented professional

Places education after experience; includes major and related minor plus strong GPA

Related Skills and Achievements

Not every résumé will have this section but the following are all employer-friendly things to include:

- second or third language(s) you speak or write
- extensive travel
- certificates or licenses you hold
- memberships in professional associations (e.g., American Society of Safety Engineers, Black Student Association, National Hispanic Business Association, Child Development Organization)
- memberships in community groups (e.g., Lions, Red Cross, Elks); list any offices you hold—recorder, secretary, fund drive chairperson

Computer Skills

Knowledge of computer hardware, software, word processing programs, and web design and search engines is extremely valuable in a job market. Note how Anthony Jones and Maria Lopez inform perspective employers about their relevant computer competencies in Figures 5.3 and 5.4.

Honors/Awards

List any civic (mayor's award, community service, cultural harmony award) and/or academic (dean's list, department awards, scholarships, grants, honorable mentions) honors you have won. Memberships in honor societies in your major and technical/business associations also demonstrate that you are professionally accomplished and active.

References

You can simply say that you will provide references on request or you can list the names, titles, e-mail and street addresses, and telephone numbers of no more than three or four individuals, as María Lopez did in Figure 5.4. Be sure to obtain their permission first. List your references only when they are well known in the community or belong to the same profession in which you are seeking employment—you profit from your association with a recognizable name or title.

Asking your boss can be tricky. If your current employer knows that your education is preparing you for another profession, ask her or him. However, if you are employed and are looking for professional advancement or a better salary elsewhere, you may not want your present employer to know you are searching for another job until you become a leading candidate.

In this section of your résumé, you may also indicate that a portfolio of your work is available for review, as Anthony Jones did in Figure 5.3.

Organizing Your Résumé

There are two primary ways to organize your résumé: chronologically or by function or skill area.

Chronologically

The résumés in Figures 5.3 and 5.4 are organized chronologically. Information about the job applicants is listed year by year under two main categories—education and experience. This is the traditional way to organize a résumé. It is straightforward and easy-to-read, and employers find it acceptable. The chronological sequence works especially well when you can show a clear continuity toward progress in your career through your job(s) and in schoolwork or when you want to apply for a similar job with another company.

A chronological résumé is appropriate for students who want to emphasize recent educational achievements or when there are no major gaps in your career.

By Function or Skill Area

Depending on your experiences and accomplishments, you might organize your résumé according to function or skill area. According to this plan, you would *not* list your information chronologically in the categories "Experience" and "Education." Instead, you would sort your achievements and abilities—whether from course work, jobs, extracurricular activities, or technical skills—into two to four key skill areas, such as "Sales," "Public Relations," "Training," "Management," "Research," "Technical Capabilities," "Counseling," "Group Leadership," "Communications," "Network Operations," "Customer Service," "Working with People," "Multicultural Experiences," "Computer Skills," "Problem-Solving Skills."

Under each area you would list three to five points illustrating your achievements in that area. Skills or functional résumés are often called **bullet résumés** because they itemize the candidate's main strengths in bulleted lists. Some employers prefer the bullet résumé because they can skim the candidate's list of qualifications in a few seconds.

Note Dora Cooper Bolger's profitable use of a functional résumé format in Figure 5.5. She was out of school because of family commitments, yet she uses the experiences she acquired during those years to her advantage in her résumé organized by "Skills, Responsibilities, Experiences." She successfully translates her many accomplishments in managing a home and working on charitable and community projects into marketable skills of great interest to a prospective employer, and no gap of ten years interrupts a work experience list.

Who Should Use a Functional Résumé?

The following individuals would probably benefit from organizing their résumés by function instead of chronologically:

- nontraditional students who have diverse job experiences
- people who are changing their profession because of downsizing or seeking new professional opportunities
- individuals who have changed jobs frequently over the last five to ten years
- ex-military personnel re-entering the civilian marketplace upon conclusion of their military service

You might want to prepare two different versions of your résumé—one functional and one chronological—to see which sells your talents better. Don't hesitate to seek the advice of your instructor or placement counselor about which one will work best for you.

The Online Résumé

In addition to preparing a hard copy of your résumé, expect to create one (or more) to be posted online. An online résumé will give you the widest possible exposure to attract prospective employers.

The online résumé can be delivered in several ways. You can post it on a database service such as **hotjobs.yahoo.com** or **monster.com** or on databases sponsored by professional societies, such as ACM in Figure 5.2. Expect to send your online résumé directly to an employer's website as well, where it can be scanned, indexed, and even stored. Or you can simply send your résumé to a prospective employer via e-mail. You can also post your résumé on your own website. But job counselors caution that posting your résumé on your own website may entail providing more information than the prospective employer needs, or giving out too much personal information.

Prospective employers may receive hundreds of résumés in response to a single job posting. Because of such volume, employers will often scan résumés very quickly to make the first cut. You need to prepare your online résumé so that it captures the attention of the maximum number of employers. To accomplish this, design your online résumé so that it can be read clearly and quickly and so that it contains keywords that match your prospective employer's needs (see page 149).

While an online résumé contains basically the same information as the print résumé, as discussed on pages 143–146, the design of an online résumé is very different. Don't think that you can just paste your hard copy résumé into an e-mail or send it to a database or employer's web portal. You will first have to format it in electronically scannable form. Otherwise, your résumé may be blocked or garbled. Figure 5.6 shows an online version of Anthony Jones's résumé in Figure 5.3.

Formatting an Online Résumé

Here are some guidelines to help you make sure your résumé is scannable:

1. Create your résumé as an RTF, HTML, or MS Word file (the most common file formats in the world of work) rather than as a PDF, WordPerfect, or Zip file.
2. Save your résumé properly and consistently—as an ASCII plain text file. Follow the directions given by a database service or the prospective employer.
3. Avoid underlining, boldface, italics, boxes, or shadowing, which can garble the text of your résumé and make it unscannable. Use all caps instead of bold or italics for emphasis, and an asterisk (*) or a plus sign (+) in place of a bullet at the beginning of a line, as in Figure 5.6.
4. Do not use hard to read or fancy fonts like script. Instead, choose a font like Courier 12 that is clear and does not mask letters.
5. Make sure your résumé is easy to scroll. Keep in mind that reading a résumé on a screen is different from seeing it on a printed page. Use plenty of white

An online résumé.   **Figure 5.6**

Anthony H. Jones
webguru@gmail.com
Phone: (707) 555-6390

OBJECTIVE

Position as layout and Web design editor with commercial publisher

KEYWORDS

Web designer, computer graphics, Illustrator, Photoshop, QuarkXPress, fundraiser, budgets, sales, Soapscan, WriteNow, virus protection, team player

EDUCATION

Santa Rosa Junior College, A.S. degree to be awarded in June 2008. Commercial Graphics Illustration major. Digital photography minor. GPA 3.45

COMPUTER SKILLS

Excellent working knowledge of computer graphics: Illustrator, Photoshop, QuarkXPress Passport, WordPerfect v.3, Soapscan, WriteNow

EXPERIENCE

* Intern in layout and design department. Preparing page composition, importing visuals, manipulating images, McAdam Publishers, 8 Parkway Heights, Santa Rosa, CA 94211

* Salesperson; display merchandise coordinator, Buchman's Department Store, Greenview Mall, Santa Rosa

* Volunteer: designed website, brochures, and other artwork for successful fund drive, Santa Rosa Humane Society

* Web designer, graphic artist, proofreader, Thunder, student magazine

REFERENCES AVAILABLE ON REQUEST

All lines aligned flush with the left margin

Objective appears first within the body of the online résumé

Keywords (for search engine readiness) go at the top of the online résumé

All caps rather than bold, italic, fancy fonts used to highlight categories

Uses terminology appropriate for position

Chooses nouns rather than action verbs to increase employer matches

Asterisks rather than bullets mark beginning of lines

space between sections so that headings are clear and distinct from one another. Put no more than 65 characters on a line to make your résumé easy to read.

6. Design your résumé to be concise. Prospective employers will not have the time or patience to scroll through multiple screens to find information. The equivalent of one printed page is still best. Don't use a "page 2" or "continued."

7. Select the automatic wrapping feature to avoid losing characters at the end of a line but make sure you do not exceed the limit of 65 characters per line.

8. Align everything flush with the left margin. Do not indent anything or use tabs.

9. Never send your résumé as an e-mail attachment. A prospective employer will not open it because of the risk of contracting a virus and because it takes too long. Copy and paste it into the body of your e-mail message.

10. Include a short cover letter (two or three brief paragraphs, not just one or two sentences), placed before your résumé, to inform an employer that you are applying for a specific job (and name it).

11. Format and save your résumé before you send it to any database service.

12. Preview your résumé for quality control.

Making Your Online Résumé Search-Engine Ready

The most important part of your online résumé is the keywords you use. Prospective employers scan résumés to find the keywords they most want to see in the job seeker's description of his or her experience, education, and interpersonal skills. The more matches, or hits, they find between appropriate keywords on your résumé and those on their list, the better your chances are of being interviewed. List keywords at the top of your résumé, as in Figure 5.6, and also insert them throughout your résumé in appropriate places. Keywords should highlight your technical expertise, training and education, knowledge of a field, leadership ability, writing/speaking skills, sales experience, etc.

Here are a few tips to help you select appropriate keywords:

1. Include the keywords found in the employer's ad and website to increase your chances of landing an interview.

2. Do not be afraid of using shoptalk (or jargon) of your profession. An employer will expect you to be aware of current terminology.

3. Use keywords to head and connect sections or categories of your résumé. Keyword headers will help you emphasize your job strengths and make it easy for employers to scroll back to an appropriate section of your résumé.

4. Replace the action verbs found in conventional résumés on the left with keyword nouns on the right. Here are some examples:

Conventional Résumé	Online Résumé
developed supplementary brochures	ancillary development editor
wrote technical report	technical writer
performed laboratory tests	laboratory technician
responsible for managing accounts	accounts manager

| won two awards | award winner |
| assisted employees at software help desk | IT help desk specialist |

Testing, Proofing, and Sending Your Online Résumé

1. Test your formatting. Send your résumé to a friend who does not use your Internet provider to be sure your file is readable and formatted correctly.
2. Proofread carefully just as you would a hardcopy résumé.
3. Don't just put *résumé* as the subject of your file to an employer. List the title/number or code of the position for which you are applying.
4. Simply posting your résumé online is not enough. Also send a scannable hard copy and a cover letter to prospective employers. But do not fold or staple your résumé. Put it, along with your cover letter, in a large envelope.
5. Always keep a log of where you have sent your résumé online.

Cyber-Safing Your Résumé

Whether you use a database service or post your résumé on a website, protect your identity and your current and future jobs. Be careful about revealing personal information.

- Post your résumé only on legitimate sites. Avoid those that say they will flood the market. You don't know where your résumé will end up.
- Do not put personal information in your résumé—home address, social security number, birthday, health status, or photograph.
- Use an anonymous e-mail address rather than your personal one. Consider a generic e-mail address that includes a word or phrase that identifies your area of expertise, for example, *JosieProgrammer@aol.com.* Or include an e-mail link with a built-in "mailto" on your résumé.
- Never put the names of your references or their addresses online. Simply say, "References available on request."
- Block certain readers from searching your résumé, such as your current employer or firms that you know send out spam.
- Never use your present employer's company name or business e-mail address.
- The online résumé is *not* the place to allow readers access to a portfolio of your work. Send your portfolio only to employers who request it.

■ Letters of Application

Along with your résumé, you must send your prospective employer a letter of application, one of the most important pieces of correspondence you may ever write. Its goal is to get you an interview and ultimately the job. Letters you write in applying for jobs should be personable, professional, and persuasive—the three **P**'s. Knowing how the letter of application and résumé work together and how they differ can give you a better idea of how to compose your letter.

Chapter 5 Additional Activities, located at **college.cengage .com/pic/ kolinconcise2e**, include a global-focused activity titled "Applying for an International Job."

How Application Letters and Résumés Differ

The résumé is a compilation of facts—a record of dates, your important achievements, names, places, education, and jobs. As noted earlier, you may prepare several different résumés depending on your experience and the job market.

Your letter of application, however, is much more personal. Because you must write a new, original letter to each prospective employer, you may write (or adapt) many different letters. Each letter of application should be tailored to a specific job. It should respond precisely to the kinds of qualifications the employer seeks.

The letter of application is a sales letter that emphasizes and applies the most relevant details (of education, experience, and talents) in your résumé. In short, the résumé contains the raw material that the letter of application transforms into a finished and highly marketable product—you.

Résumé Facts to Exclude from Letters of Application

The letter of application should not simply repeat the details listed in your résumé. In fact, the following details are relevant information you list in your résumé and should *not* be restated in the letter:

- personal data, including license or certificate numbers
- specific names of courses in your major
- names and addresses of all your references

Writing the Letter of Application

The letter of application can make the difference between your getting an interview and being eliminated early from the competition. It should convince a prospective employer that you will use the experience and education listed on your résumé on the job he/she is hoping to fill. You want your letter to be placed in the "definitely interview" category. Limit your letter to one page. As you prepare your letter, use the following general guidelines.

1. Follow the standard conventions of letter writing (see Chapter 4). Print your letter on high-quality, white bond 8.5″ × 11″ paper. Proofread meticulously; a spelling error, typo, or grammatical mistake will make you look careless. Don't trust spell-check.

2. Make sure your letter looks attractive. Use wide margins and don't crowd your page. Keep your paragraphs short and readable—four or five sentences (see pp. 168–174 in Chapter 6).

3. Send your letter to a specific person. Never address an application letter "To Whom It May Concern," "Dear Sir or Madam," or "Director of Human Resources." Get an individual's name by double-checking the company's website or by calling the company's headquarters and be sure to verify the spelling of the person's name and his or her title.

4. Don't forget the "you attitude" (see Chapter 4, pp. 89–94). See yourself as an employer sees you. Focus on how your qualifications meet the employer's needs, not the other way around. Employers are not impressed by vain boasts ("I am the most efficient and effective safety engineer"). Convince readers that you will be a valuable addition to their organization. Cite specific skills, duties, and accomplishments.

5. **Don't be tempted to send out your first draft.** Write and rewrite your letter of application until you are convinced it presents you in the best possible light. Getting the job may depend on it. A first or even second draft rarely sells your abilities as well as a third, fourth, or even fifth revision does.

The sections that follow give you some suggestions on how to prepare the various parts of an application letter successfully.

Your Opening Paragraph

The first paragraph of your letter of application is your introduction. It must get your reader's attention by answering three questions:

1. Why are you writing?
2. Where or how did you learn of the vacancy or the company or the job?
3. What is your most important qualification for the job?

Begin your letter by stating directly that you are writing to apply for a job. Don't say that you "want to apply for the job"; such an opening raises the question, "Why don't you, then?"

Avoid an unconventional or arrogant opening: "Are you looking for a dynamic, young, and talented photographer?" Do not begin with a question; be more positive and professional.

If you learned about the job through a website or journal, make sure you italicize or underscore its title.

> I am applying for the food-service manager position you advertised in the May 10 edition of the *Los Angeles Times* online.

Since many companies announce positions on the Internet, check there first to see if their position is listed online, as Anthony Jones did (Figure 5.7).

If you learned of the job from a professor, a friend, or an employee at the firm, state so. Take advantage of a personal contact who is confident that you are qualified for the position, as María Lopez (Figure 5.8, p. 155) and Dora Cooper Bolger (Figure 5.9, p. 157) did. But first confirm that your contact gives you permission to use his or her name.

The Body of Your Letter

The body of your letter, comprising one or two paragraphs, provides the evidence from your résumé to prove you are qualified for the job. You might want to spend one paragraph on your education and one on your experience or combine your accomplishments into one paragraph.

Follow these guidelines for the body of your letter:

1. **Keep your paragraphs short and readable—four or five sentences.** Avoid long, complex sentences. Use the active voice to emphasize yourself as a doer. Review the action verbs in Table 5.1 and use keywords found in the employer's ad.

Figure 5.7   Letter of application from Anthony Jones, a recent graduate with little job experience.

Clear, easy-to-follow letterhead

WEBSITE DEVELOPER
DESIGNER
GRAPHIC ARTIST

Anthony H. Jones
73 Allenwood Boulevard
Santa Rosa, California 95401-1074
707-555-6390
ajones@plat.com
www.plat.com/users/ajones/resume.html

May 24, 2008

Ms. Jocelyn Nogasaki
Human Resources Manager
Megalith Publishing Company
1001 Heathcliff Row
San Francisco, CA 94123-7707

Dear Ms. Nogasaki:

Identifies position and source of ad

I am applying for the layout editor position advertised on your website, which I accessed on 14 May. Early next month, I will receive an A.S. degree in commercial graphics illustration from Santa Rosa Junior College.

Validates necessary education and applies it directly to employer's business

With a special interest in the publishing industry, I have successfully completed more than 40 credit hours in courses directly related to layout design, where I acquired experience using QuarkXPress as well as Illustrator and Photoshop. You might like to know that many Megalith publications were used as models of design and layout in my graphics communications and digital photography classes.

Convincingly cites related job experience

My studies have also led to practical experience at McAdam Publishers as part of my Santa Rosa internship program. While working at McAdam, I was responsible for assisting the design department in page composition and importing images. Other related experience I have had includes creating a website for the Santa Rosa Humane Society. As you will note on the enclosed résumé, I have also had on-the-job training in displaying merchandise at Buchman's Department Store.

Refers to résumé

Asks for an interview and gives contact information

I would appreciate the opportunity to discuss with you my qualifications in graphic design. My phone number is 707.555.6390. After June 12, I will be available for an interview at any time that is convenient for you.

Sincerely yours,

Anthony H. Jones

Anthony H. Jones
Encl. Résumé

Letter of application from María Lopez, a recent graduate with some job experience.   Figure 5.8

1725 Brooke Street
Miami, FL 32701-2121
mlopez@eagle.com   cell: 305-555-3429

May 15, 2009

Dr. Marvin Henrady
Suite 34
Medical/Dental Plaza
839 Causeway Drive
Miami, FL 32706-2468

Dear Dr. Henrady:

Mr. Mitchell Pelbourne, my clinical instructor at Miami-Dade Community College, informs me you are looking for a dental hygienist to work in your northside office. My education and experience qualify me for that position. This month I
will graduate with an A.S. degree in the dental hygienist program, and I will take the American Dental Assistants' Examination in early June.

I have successfully completed all course work and clinical programs in oral hygiene, anatomy, and prophylaxis techniques. During my clinical training, I received intensive practical instruction from several local dentists, including Dr. Tia Gutiérrez. Since your northside office specializes in pedodontal care, you might find the subject of my major project—proper nutrition and dental care for preschoolers—especially relevant.

I have related job experience in working with children in a health care setting. For over two years, I was a unit clerk on the pediatric unit at St. Francis Hospital, and my experience in greeting patients, transcribing orders, and assisting the nursing staff would be valuable to you in your office. An additional job strength I would bring to your office is my bilingual (Spanish/English) communication skills. You will find more detailed information about my qualifications in the enclosed résumé.

I would welcome the opportunity to talk with you about the position and my interest in pedodontics. I am available for an interview any time after 2:30 until June 11. After that date, I could come to your office any time at your convenience.

Sincerely yours,

María H. Lopez

María H. Lopez

Encl. Résumé

Begins with personal contact

Verifies she will have necessary licensure

Links training to job respon-sibilities; demonstrates knowledge of employer's office

Relates previ-ous experience to employer's needs; refers to résumé

Ends with a polite request for an interview

2. **Don't begin each sentence with "I."** Vary your sentence structure. Write reader-centered sentences, even those beginning with "I."

3. **Concentrate on seeing yourself as your employer sees you.** Prove that you can help an employer's sales and service, promote an organization's mission and goals, and be a reliable team player.

4. **Highlight your qualifications by expressing accomplishments in numbers or percentages.** Tell your reader exactly how your schoolwork and job experience qualify you to perform and advance in the job advertised. Don't simply say you are a great salesperson. Stress that you increased the sales volume in your department by 20 percent within the last 6 months.

5. **Mention you are enclosing your résumé.** Make sure you attach it.

Education.   Recent graduates with little work experience will, of course, spend more time on their education. But regardless, emphasize why and how your most significant educational accomplishments—course work, degrees, training—are relevant for the particular job. Employers want to know which specific skills from your education translate into benefits for their company.

Simply saying you will graduate with a degree in criminal justice does not explain how you, unlike all the other graduates of such programs, are best suited for a particular job. But when you indicate that you have completed 36 credit hours in software security and have another 12 credit hours in global business marketing, you show exactly how and where you are qualified. Note how Anthony Jones in Figure 5.7 and María Lopez in Figure 5.8 establish their educational qualifications with specific details about their training.

Experience.   After you discuss your educational qualifications, turn to your job experience. But if your experience is your most valuable and extensive qualification for the job, put it before education. If you are switching careers or returning to a career after years away from the work force, start the body of your letter with your experience or your community and civic service, as Dora Cooper Bolger does in Figure 5.9. Her volunteer work convincingly demonstrates she has the organizational and communication skills her prospective employer seeks. Never minimize such contributions.

Relate Your Experience/Education to the Job.   Persuasively show a prospective employer how your previous accomplishments have prepared you for future success on the job. Relate your course work in computer science to being an efficient programmer. Indicate how your summer jobs for a local park district reinforced your exemplary skills in customer service. Connect your background to the prospective employer's company. Any homework you can do about the company's history, goals, or structure (see p. 159) will pay off.

- By citing Megalith publications as a model in his courses, Anthony Jones in Figure 5.7 stresses he is ready to start successfully from the first day on the job.

Letter of application from Dora Cooper Bolger, a recent graduate with community and civic experience.

Figure 5.9

Dora Cooper Bolger
1215 Lakeview Avenue
Westhampton, MI 46532
Voice: 616-555-4772   **Cell:** 616-555-4773   **dbolger@aol.com**

February 9, 2009

Dr. Lindsay Bafaloukos
Tanselle Mental Health Agency
4400 West Gallagher Drive
Tanselle, MI 46932-3106

Dear Dr. Bafaloukos:

At a recent meeting of the County Services Council, a member of your staff, George Azmar, told me that you will be hiring a public affairs coordinator. Because of my extensive experience in and commitment to community affairs, I would appreciate your considering me for this opening. I expect to receive my B.S. in Public Administration from Mid-Michigan College later this year.

For the past ten years, I have organized community groups with outreach programs similar to Tanselle's. I have held administrative positions in the PTA and the Foster Parents' Association and was president of Teens in Trouble, a volunteer group providing assistance to dysfunctional teens. My responsibilities with Teens have included coordinating our activities with various school programs, scheduling tutorials, and representing the organization before local and state agencies. I have been commended for my organizational and communication skills. My eighteen presentations on foster home care and Teens in Trouble demonstrate that I am an effective speaker, a skill your agency would find valuable.

Because of my work at Mid-Michigan and in Teens and Foster Parents, I have the practical experience in communication and psychology to promote Tanselle's goals. The enclosed résumé provides details of my experience and education.

I would enjoy discussing my work with Teens and the other organizations with you. I am available for an interview any day after 11:00 a.m. Thank you for reaching me at the phone numbers or e-mail address above.

Sincerely yours,

Dora Cooper Bolger

Dora Cooper Bolger

Encl. Résumé

Begins with contact made at professional meeting, highlighting her interests

Relates proven past successes to employer's needs; gives concrete examples of her skills

Encourages reader to see her as best-prepared candidate

Requests interview and gives contact information

Includes résumé

- Note how María Lopez in Figure 5.8 links her work on a pediatric unit of a hospital to Dr. Henrady's specialty in pedodontics.
- Dora Cooper Bolger in Figure 5.9 likewise shows her prospective employer that she is familiar with and can contribute to Tanselle's programs in community mental health through her volunteer work and public speaking experience.

Closing

Keep your closing paragraph short—about two or three sentences—but be sure it fulfills the following three important functions:

1. emphasizes once again briefly your major qualifications
2. asks for an interview or a phone call
3. indicates when you are available for an interview

End gracefully and professionally. Be straightforward. Don't leave the reader with a single weak, vague sentence: "I would like to have an interview at your convenience." That does nothing to sell you. Say that you would appreciate talking with the employer further to discuss your qualifications. Then mention your chief talent. You might also express your willingness to relocate if the job requires it.

After indicating your interest in the job, give the times you are available for an interview and specifically tell the reader where you can be reached. If you are going to a professional meeting that the employer might also attend, or if you are visiting the employer's city soon, say so.

The following samples show how *not* to close your letter and why not.

Pushy:	I would like to set up an interview with you. Please phone me to arrange a convenient time. (That's the employer's prerogative, not yours.)
Too Informal:	I do not live far from your office. Let's meet for coffee sometime next week. (Say instead that since you live nearby, you will be available for an interview.)
Introduces New Subject:	I would like to discuss other qualifications you have in mind for the job. (How do you know what the interviewer might have in mind?)

Note how the closing paragraphs in Figures 5.7 through 5.9 avoid these errors.

▮ Going to an Interview

There are various ways for a prospective employer to conduct an interview. It might be a one-to-one meeting—you and the interviewer. Or you may visit with a group of individuals who are trying to decide if you would fit in. Or you might have your interview over the telephone or through a videoconference. An interview can last thirty minutes or take all day.

Preparing for the Interview

Before you go to the interview, be as prepared as you can possibly be by doing the following:

1. Do your homework about the company. Click on **About Us** on the company's website and obtain a copy of its annual stockholders' report, if possible, to find out who founded the company, who the current CEO is, what its chief products/services are, how many years it has been in business, how many people it employs, where its offices/ plants are located, and who are its major clients and competitors.

2. Review the job description carefully (and bring a copy of it with you to review prior to the interview) so that you are completely clear about what the job entails.

3. Prepare a one-minute summary of your chief qualifications for the job, as the interviewer(s) will most likely ask you to summarize your qualifications.

4. Practice your interview skills with a friend or job skills counselor. Be sure that this person asks tough, relevant questions so that you can practice answering the types of questions you will get at the interview.

5. Brush up on business etiquette, including such important considerations as being on time, remembering the name of the interviewer and other people you will meet, and being polite and respectful especially of other cultures.

6. Bring three or four extra copies of your résumé with you, as well as a note pad and pen, or a Blackberry or PDA (if you have either) to jot down essential details. Leave your laptop at home.

7. Bring your photo ID, social security card (or work visa, if you are not a U.S. citizen), and any other licenses/certificates you may be asked to present to a Human Resources Department.

Questions to Expect

The following questions are typical of those you can expect from interviewers, with advice on how to answer them.

- **Why do you want to work for us?** (Recall any job goals you have and apply them specifically to the job under discussion.)
- **What qualifications do you have for the job?** (Point to educational achievements and relevant work experience, especially computer skills.)
- **What could you possibly offer us that other candidates do not have?** (Say "enthusiasm," being a team player, problem-solving abilities, leadership skills.)
- **Why did you attend this school?** (Be honest—location, costs, programs.)
- **Why did you major in "X"?** (Do not simply say financial benefits; concentrate on both practical and professional benefits. Be able to state career objectives.)
- **Why did you get a grade of "C" in a course?** (Don't say that you could have done better if you'd tried. Explain what the trouble was and mention that you corrected it in a subsequent course in which you earned a B or an A.)

- **What extracurricular activities did you participate in while in high school or college?** (Indicate any responsibilities you had—handling money, writing reports, designing web documents, coordinating events. If you were unable to participate in such activities, tell the interviewer that a part-time job, community or church activities, or commuting prevented your participating. Such answers sound better than saying that you did not like sports or clubs in school.)
- **Did you learn as much as you wanted from your course work?** (This is a loaded question. Indicate that you learned a great deal but now look forward to the opportunity to gain more practical skill, to put into practice the principles you have learned; say that you will never be through learning about your major.)
- **What is your greatest strength?** (Say being a team player, sensitive to others' needs, cooperation, willingness to learn, ability to grasp difficult concepts easily, planning and organizing tasks, managing time or money, taking criticism easily, and profiting from criticism.)
- **What is your greatest shortcoming?** (Be honest here and mention it, but then turn to ways in which you are improving. Don't say something deadly like, "I can never seem to finish what I start" or "I hate being criticized." You should neither dwell on your weaknesses nor keep silent about them. Saying "None" to this kind of question is as inadvisable as rattling off a list of faults.)
- **How do you handle conflict with a co-worker, boss, customer?** (Stress your ability to be courteous and honest and to work toward a productive resolution. State that you avoid language, tone of voice, or gestures that interfere with healthy dialogue. Express your willingness to collaborate.)
- **Why did you leave your last job?** ("I returned to school full-time" or "I moved from Jackson to Springfield." "I had greater career possibilities." *Never attack your previous employer.* That only makes you look bad.)
- **Why would you leave your current job?** (Again, never attack an individual or an organization. Say your current job has prepared you for the position you are now applying for. Emphasize your desire to work for a new company because of its goals, work environment, opportunities, and challenges.)

What Do I Say About Salary?

Again, do your homework. Find out what the salary range is for your professional level in your area. Consult the U.S. Bureau of Labor Statistics at *http://www.bls.gov.oco* as well as *http://www.salary.co*. Ask your instructors or individuals you know who work for the company. If the issue of salary comes up, ask if the company has established a salary range for the position and where you stand in relationship to that range. However, since many companies set fixed salaries for entry-level positions, it may be unwise to try to negotiate.

If you are asked about what salary you expect for the job, do not give an exact figure. You may undercut yourself if the employer has a higher figure in mind. By doing your homework on salary ranges, you will have a better feel for the market if and when the employer does mention salary.

Factor other things into your job equation—insurance, day care, housing, uniform/clothing allowances, product or service discounts, retirement plans, and tuition reimbursement. See Figure 1.1 for an example of such a job perk.

Ten Interview Dos and Don'ts

Keep in mind these other interview dos and don'ts.

1. Be on time. In fact, show up about fifteen minutes earlier in case the interviewer wants you to complete some forms.
2. Go to the interview alone. Turn your cell phone off!
3. Dress appropriately for the occasion. Be well groomed. Never wear blue jeans. **Men:** Wear a suit and tie. **Women:** Wear a suit (pants or skirt) or equally businesslike attire.
4. Be careful about tattoos. Job counselors warn that visible tattoos can hurt a job seeker's chance for success.
5. Greet the interviewer with a friendly and firm, not vicelike, handshake. Thank your interviewer(s) for inviting you.
6. Don't sit down before the interviewer does. Wait for him or her to invite you to sit and to indicate where.
7. Be enthusiastic but not arrogant. Speak slowly and distinctly; do not nervously hurry to finish your sentences and never interrupt or finish an interviewer's sentences. Don't talk too much. Avoid one- or two-word answers, which sound unfriendly or unprepared. Do not use slang (e.g., "Right on," "Way to go," "You go, girl").
8. Refrain from chewing gum; clicking a ballpoint pen; fidgeting; twirling your hair; or tapping your foot against the floor, a chair, or a desk.
9. Maintain eye contact with the interviewer; do not sheepishly stare at the floor or the desk. Body language is equally important. For instance, don't fold your arms—a signal that you are closed to the interviewer's suggestions and comments. Sit up straight; do not slouch.
10. When the interview is over, thank the interviewer for considering you for the job and say you look forward to hearing from him or her.

The Follow-Up Letter

Within a week after the interview, it is wise to send a follow-up letter thanking the interviewer for his or her time and interest in you. In your letter, you can re-emphasize your qualifications for the job by showing how they apply to conditions described by the interviewer; you might also ask for further information to show your interest in the job and the employer. A sample follow-up letter appears in Figure 5.10.

Figure 5.10   A follow-up letter.

2739 East Street
Latrobe, PA 17042-0312

610-555-6373
mlb@springboard.com

September 20, 2008

Mr. Jack Fukurai
Manager of Human Resources
Global Tech
1334 Ridge Road N.E.
Pittsburgh, PA 17122-3107

Dear Mr. Fukurai:

I enjoyed talking with you last Wednesday and learning more about the security officer position available at Global Tech. It was especially helpful to take a tour of the plant's north gate section to see the challenges it presents for the security officer stationed there.

As you noted at the interview, my training in surveillance electronics has prepared me to operate the state of the art equipment Global Tech has installed at the north gate. I was grateful to Ms. Turner for taking time to demonstrate the equipment.

I am looking forward to receiving the handbook about employee services. Would you kindly e-mail me a copy of the newsletter from last year that introduced the new security system to Global Tech employees?

Thank you for considering me for the position and for the hospitality you showed me. I look forward to hearing from you. After my visit last week, I know that Global Tech would be an excellent place to work.

Sincerely yours,

Marcia Le Borde

Marcia Le Borde

Revision Checklist

- ☐ Inventoried my strengths carefully to prepare résumé.
- ☐ Restricted the types of job(s) for which I am qualified.
- ☐ Requested letters from professors, employers, and community officials.
- ☐ Identified places where relevant jobs are advertised.
- ☐ Networked with instructors, friends, relatives, clergy, and individuals who work for the companies I want to join that I am in the job market.
- ☐ Checked with state employment office and relevant government agencies.
- ☐ Researched companies I am interested in—on the Internet, through news stories, etc.
- ☐ Wrote a focused and persuasive career objective.
- ☐ Determined the most beneficial format of résumé to use—chronological, functional, or both.
- ☐ Prepared an electronic résumé to send online; made sure it was formatted properly.
- ☐ Cyber-safed résumé to protect personal information.
- ☐ Researched companies I was interested in through news stories on the Internet.
- ☐ Made résumé attractive and easy to read with logical and persuasive headings and keywords.
- ☐ Made sure résumé contains neither too much nor too little information.
- ☐ Proofread résumé to ensure everything is correct, consistent, and accurate.
- ☐ Wrote letter of application that shows how my specific skills and background apply to and meet an employer's exact needs.
- ☐ Prepared for interview.
- ☐ Sent prospective employer a follow-up letter within a few days after interview to show interest in position.

Exercises

1. Using at least four different sources, including the Internet, compile a list of ten employers for whom you would like to work. Get their names, street and e-mail addresses, phone numbers, and the names of the managers or human resources officers. Then select one company and write a profile about it—locations, services, kinds of products or services offered, number of employees, clients served, types of schedules used, and other pertinent facts.

2. Which of the following would belong on your résumé? Which would not belong? Which would be found on a hardcopy but not online version of your résumé? Why?

Additional Activities related to how to get a job are located at **college .cengage .com/pic/ kolinconcise2e**.

a. student ID number
b. social security number
c. the ZIP codes of your references' addresses
d. a list of all your English courses in college
e. section numbers of the courses in your major
f. statement that you are recently divorced
g. subscriptions to journals in your field
h. the titles of any stories or poems you published in a high school literary magazine or newspaper
i. your GPA
j. foreign languages you studied
k. years you attended college
l. the date you were discharged from the service
m. names of the neighbors you are using as references
n. your religion
o. job titles you held
p. your summer job washing dishes
q. your telephone number
r. the reason you changed schools
s. your current status with the National Guard
t. the URL of your website
u. your volunteer work for the Red Cross
v. hours a week you spend reading science fiction
w. the title of your last term paper in your major
x. the name of the agency or business where you worked last

3. Indicate what is wrong with the following career objectives and rewrite them to make them more precise and professional.
a. Job in a dentist's office.
b. Position with a safety emphasis.
c. Desire growth position in a large department store.
d. Am looking for entry position in health sciences with an emphasis on caring for older people.
e. Position in sales with fast promotion rate.
f. Want a job working with semiconductor circuits.
g. I would like a position in fashion, especially one working with modern fashion.
h. Desire a good-paying job, hours: 8–4:30, with double pay for overtime. Would like to stay in the Omaha area.
i. Insurance work.
j. Working with computers.
k. Personal secretary.
l. Job with preschoolers.
m. Full-time position with hospitality chain.
n. I want a career in nursing.
o. Police work, particularly in suburb of large city.

    p. A job that lets me be me.

    q. Desire fun job selling cosmetics.

    r. Any position for a qualified dietitian.

    s. Although I have not made up my mind about which area of forestry I shall go into, I am looking for a job that offers me training and rewards based on my potential.

4. As part of a team or on your own, revise the following poor hardcopy résumé to make it more precise and persuasive. Include additional details where necessary and exclude any details that would hurt the job seeker's chances. Also correct any inconsistencies.

RÉSUMÉ OF

Powell T. Harrison
8604 So. Kirkpatrick St.
Ardville, Ohio
345 37 8760
614 234 4587
harrison@gem.com

PERSONAL	Confidential
CAREER OBJECTIVE	Seek good paying position with progressive Sunbelt company.
EDUCATION	
2006–2008	Will receive degree from Central Tech. Institute in Arch. St. Earned high average last semester. Took necessary courses for major; interested in systems, plans, and design development.
2003–2007	Attended Ardville High School, Ardville, OH; took all courses required. Served on several student committees.
EXPERIENCE	None, except for numerous part-time jobs and student apprenticeship in the Ardville area. As part of student app. worked with local firm for two months.
HOBBIES	Surfing the Net, playing Final Fantasy. Member of Junior Achievement.
REFERENCES	Please write for names and addresses.

---

5. Explain why the following letter of application is ineffective. Rewrite it to make it more precise and appropriate.

Apartment 32
Jeggler Drive
Talcott, Arizona

Monday

Grandt Corporation
Production Supervisor
Capital City, Arizona

Dear Sir:

I am writing to ask you if your company will consider me
for the position you announced in the newspaper yesterday.
I believe that with my education (I have an associate
degree) and experience (I have worked four years as a
freight supervisor), I could fill your job.

My schoolwork was done at two junior colleges, and I took
more than enough courses in business management and modern
technology. In fact, here is a list of some of my courses:
Supervision, Materials Management, Work Experience in
Management, Business Machines, Safety Tactics,
Introduction to Packaging, Art Design, Modern Business
Principles, and Small Business Management. In addition, I
have worked as a loading dock supervisor for the last two
years, and before that I worked in the military in the
Quartermaster Corps.

Please let me know if you are interested in me. I would
like to have an interview with you at the earliest
possible date, since there are some other firms also
interested in me, too.

Eagerly yours,

George D. Milhous

---

6. From the Sunday edition of your local newspaper or from one of the other sources discussed in the "Looking in the Right Places for a Job" section (pp. 134–136), find notices for two or three jobs you believe you are qualified to fill and then write a letter of application for one of them.

7. Write a chronologically organized résumé to accompany the letter you wrote for Exercise 6.

8. Write a functional résumé for your application letter in Exercise 6.

9. Prepare an online version of the résumé in either Exercise 7 or 8. Use persuasive keywords.

10. Bring the résumé you prepared for Exercise 7 or 8 to class to be critiqued by a collaborative writing team. After your résumé is reviewed, revise it. Write an e-mail to your instructor on the revision you made and why it will help you in your job search.

11. Prepare a hard-copy portfolio suitable to send to an employer or to present at a job interview. Collect appropriate documents (copies of reports written for a class/job, graphics, awards, degrees/certificates, etc.) and arrange them in the most relevant order. Provide a brief (1–2 sentence) annotation for each.

12. Write a letter to a local business inquiring about summer employment. Indicate that you can work for only one summer and that you will be returning to school by September 1.

CHAPTER 6

Designing Successful Documents and Visuals

To expand your understanding of designing documents and visuals, take advantage of the Web Links, Additional Activities, and ACE Self-Tests at **college.cengage .com/pic/ kolinconcise2e**.

As we have seen, to be effective in the world of work you have to write clearly and concisely, but the success of your document depends as much on how it looks as on what it says. You will be expected to design professional looking memos, letters, instructions, and reports as well as to create appropriate visuals to support these documents. This chapter gives you practical advice for making your work more reader-friendly and visually appealing. It also surveys the kinds of visuals you will encounter most frequently and shows you how to read, construct, and write about them.

In designing documents, you need to project a positive, professional image of yourself, your company, your product or service. A report filled with nothing but thick unbroken, long paragraphs, with no visual clues to break them up or to make information stand out, is sure to intimidate readers and turn them away. They will conclude that your work is too complex and not worth their effort or time. Your company, too, will win or lose points because of your design choices. A visually appealing document will enhance a company's reputation and improve its sales. A poorly designed one will not. Consumers will think your firm is inflexible, difficult to do business with, and uncaring about specific problems customers may have over a policy or a set of instructions.

Make your documents look user-friendly by signaling to your audience that your message is

- easy to read
- easy to follow and understand
- easy to recall

You can do all this by breaking material into smaller units that are visually appealing. You can help readers find key points at a glance through **chunking** (using smaller paragraphs) and using lists, boldface type, and bullets. That way they can find your ideas easily the first time through or on a second reading if they have to double back to check or verify a point.

Organizing Information Visually

Take a quick look at Figures 6.1 and 6.2 (pp. 170–173). The same information is contained in each figure. Which appeals to you more? Which do you think would be easier to read? Which is better designed? As the two figures show, design or layout plays a crucial role in an audience's overall acceptance of your work.

Figure 6.2 has the positive design characteristics (on the left) while Figure 6.1 has the unfavorable ones (on the right).

▪ visual appeal	▪ crowded
▪ logical organization	▪ disorganized
▪ clarity	▪ hard to follow
▪ accessibility	▪ difficult to read
▪ functionality	▪ uneven
▪ relevance	▪ inconsistent

Study the annotations to these figures to see how the errors in Figure 6.1 are corrected in Figure 6.2. By modeling your documents after the document in Figure 6.2, you can guarantee that your message will be well received.

The ABCs of Print Document Design

The basic elements of effective document design are

- page layout
- typography, or type design, including using color
- graphics, or visuals

The proper arrangement and balance of type, white space, and graphics involve the same level of preparation that you would spend on your research, drafting, revising, and editing. Just as you research your information, you have to research and experiment in order to adopt the most effective design for your document.

Page Layout

Each of your pages needs to coordinate space and text pleasingly. Too much or too little of one or the other can jeopardize the reader's acceptance of your message. Pay attention to the following elements.

1. White space. White space, which refers to the page's open areas that are free of text and visuals, can help you increase the impact of your message. Skimping on white space by packing too much print on the page distracts the reader from the message you really want to convey. White space can entice, comfort, and appeal to the reader's "psychology of space" by

- attracting the reader's attention
- assuring the reader that information is presented logically
- announcing that information is easy to follow
- assisting the reader to organize information visually

Again, compare Figures 6.1 and 6.2. Which document was designed by someone who understands the importance of white space?

Figure 6.1   A poorly designed document.

No title

Single-spacing makes document difficult to read

The results for the recent cholesterol screening at our company's Health Fair were distributed to each employee last week. Many employees wanted to know more information about cholesterol in general, the different types of cholesterol, what the results mean, and the foods that are high or low in cholesterol.

We hope the information provided below will help employees better answer their questions concerning cholesterol and our cholesterol screening program.

High cholesterol, along with high blood pressure and obesity, is one of the primary risk factors that may contribute to the development of coronary heart disease and may eventually lead to a heart attack or stroke. Cholesterol is a fatty, sticky substance found in the bloodstream. Excessive amounts of the bad type of cholesterol can deposit on the walls of the heart arteries. *This deposit is called plaque and over a long period of time plaque can narrow or even block the blood flow through the arteries.*

Lack of headings makes it hard for readers to organize material

Total cholesterol is divided into three parts—LDL (low-density lipoprotein), or bad cholesterol; HDL (high-density lipoprotein), or good cholesterol; and VLDL (very low-density lipoprotein), a much smaller component of cholesterol you don't have to worry about. Bad (LDL) cholesterol forms on the walls of your arteries and can cause a lot of damage. Good cholesterol, on the other hand, functions like a sponge, mopping up cholesterol and carrying it out of the bloodstream.

Frequent font changes are confusing for readers

You should have received three cholesterol numbers. One is for your HDL, or good cholesterol, and the other is for your LDL, or bad cholesterol, reading. These two numbers are added to give you the third, or composite, level of your total cholesterol. *You are doing fine.*

As you can see, a total cholesterol reading of below 200 is considered safe. Continue what you have been doing. If your reading falls in the moderate risk range of 200–239, you need to modify your diet, get more exercise, and have your cholesterol checked again in six months. *If your reading is above 240, see your doctor.* You may need to take cholesterol-lowering medication, if your doctor prescribes it. Reducing your total cholesterol by even as little as 25% can decrease your risk of a heart attack by 50%.

Uneven presentation of numbers

Lack of margins makes document look dense and complex

The Surgeon General recommends that your LDL, or bad, cholesterol should be below 130. And your HDL, or good, cholesterol needs to be at least above 36. Ideally, the ratio between the two numbers should not be greater than 5 to 1. That is, your HDL should be at least 20% of your LDL. The higher your HDL is, the better, of course. So even if you have a high LDL reading, if your HDL is correspondingly high you will be at less risk.

One of the easiest ways to decrease your cholesterol is to modify your diet. Cholesterol is found in foods that are high in saturated fat. *Saturated fat comes from animal sources and also from certain vegetable sources.* Foods high in bad cholesterol that you should restrict, or avoid, include whole milk, red meat, eggs, cheese, butter, shrimp, oils such as palm and coconut, and avocados. Generally, food groups low in cholesterol include fruits, vegetables, and whole grains (wheat breads, oatmeal, and certain cereals), lean meats (fish, chicken), and beans.

Inconsistent use of italics and boldfacing

The goal of our cholesterol screening is to help each employee lower his or her cholesterol level and eventually reduce the risk of heart disease. **Besides the advice given above,** you can do the following: get regular aerobic exercise—bicycling, brisk walking, swimming, rowing—for at least 30 minutes 3–4 times a week. But get your doctor's approval first. *Eat foods low in cholesterol* but high in dietary fiber (beans, oatmeal, brown rice). Maintain a healthy weight for your frame to lower your body fat. Minimize stress, which can increase cholesterol. Learn relaxation techniques.

An effectively designed document with the same text as Figure 6.1. **Figure 6.2**

Cholesterol Screening

The results for the recent cholesterol screening at our company's Health Fair were distributed to each employee last week. Many employees wanted to know more information about cholesterol in general, the different types of cholesterol, what the results mean, and the foods that are high or low in cholesterol. We hope the information provided below will help employees better answer their questions concerning cholesterol and our cholesterol screening program.

Determining Risk Factors

High cholesterol, along with high blood pressure and obesity, is one of the primary risk factors that may contribute to the development of coronary heart disease and may eventually lead to a heart attack or stroke. Cholesterol is a fatty, sticky substance found in the bloodstream. Excessive amounts of the bad type of cholesterol can deposit on the walls of the heart arteries. This deposit is called **plaque** and over a long period of time plaque can narrow or even block the blood flow through the arteries.

Separating Types of Cholesterol

Total cholesterol is divided into three parts: (1) **LDL** (low-density lipoprotein), or bad cholesterol; (2) **HDL** (high-density lipoprotein), or good cholesterol; and (3) **VLDL** (very low-density lipoprotein), a much smaller component of cholesterol you don't have to worry about. Bad (LDL) cholesterol forms on the walls of your arteries and can cause a lot of damage. Good cholesterol, on the other hand, functions like a sponge, mopping up cholesterol and carrying it out of the bloodstream.

Title clearly set apart from text by caps, larger font, and bold-facing

Text is double-spaced with generous margins, making it more readable

Text does not look cluttered

Headings divide material into easy-to-follow units for readers

Consistent use of one font

Only key terms are boldfaced

Types of cholesterol are clearly labeled with numbers

Continued

Figure 6.2   (Continued)

Understanding Your Cholesterol Results

You should have received three cholesterol numbers. One is for your **HDL** (or good cholesterol) and the other is for your **LDL** (or bad cholesterol) reading. These two numbers are added to give you the third, or composite, level of your total cholesterol.

Cholesterol levels can be classified as follows:

Minimal Risk	Moderate Risk	High Risk
below 200	200–239	above 240

As you can see, a total cholesterol reading of below 200 is considered safe. You are doing fine. Continue what you have been doing. If your reading falls in the moderate risk range of 200–239, you need to modify your diet, get more exercise, and have your cholesterol checked again in six months. If your reading is above 240, see your doctor. You may need to take cholesterol-lowering medication, if your doctor prescribes it. Reducing your total cholesterol by even as little as 25% can decrease your risk of a heart attack by 50%.

Relationship Between Bad and Good Cholesterol

The Surgeon General recommends that your **LDL**, or bad, cholesterol should be below 130. And your **HDL**, or good, cholesterol needs to be at least above 36. Ideally, the ratio between the two numbers should not be greater than 5 to 1. That is, your **HDL** should be at least 20% of your **LDL**. The higher your **HDL** is, the better, of course. So even if you have a high **LDL** reading, if your **HDL** is correspondingly high you will be at less risk.

Recognizing Food Sources of Cholesterol

One of the easiest ways to decrease your cholesterol is to modify your diet. Cholesterol is found in foods that are high in saturated fat. Saturated fat comes

2

Continued

(Continued)   **Figure 6.2**

from animal sources and also from certain vegetable sources. Foods high in bad cholesterol that you should restrict include:

1. whole milk
2. red meat
3. eggs
4. cheese
5. butter
6. shrimp
7. oils such as palm and coconut
8. avocados

A double-spaced, numbered list helps readers identify foods with bad cholesterol

Generally, food groups low in cholesterol include fruits, vegetables, and whole grains (wheat breads, oatmeal, and certain cereals), lean meats (fish, chicken), and beans.

Easy-to-follow examples of foods low in cholesterol

Realizing It Is Up to You

The goals of our cholesterol screening program are to help each employee lower his or her cholesterol level and eventually reduce the risk of heart disease. Besides the advice given above, you can do the following:

Heading signals conclusion

- Get regular aerobic exercise—bicycling, brisk walking, swimming, rowing—for at least 30 minutes 3–4 times a week. But get your doctor's approval first.
- Eat foods low in cholesterol but high in dietary fiber (beans, oatmeal, and brown rice).
- Maintain a healthy weight for your frame to lower your body fat.
- Minimize stress, which can increase cholesterol. Learn relaxation techniques.

Bulleted list serves as both summary and plan for future action

3

Thanks to Sgt. Mannie E. Hall of the U.S. Army for creating this document.

2. Margins. Use wide margins, usually 1 to 1½ inches, to "frame" your document with white space surrounding text and visuals. Margins prevent your document from looking cluttered or overcrowded. If your document requires binding, you may have to leave a wider left margin (2 inches).

3. Line length. Most readers find a text line of 10 to 14 words, or 50 to 70 characters (depending on the type size you choose), comfortable and enjoyable to read. Excessively long lines that bump into the margins signal that your work is difficult to read. In the example here, note how the extra-long lines unsettle your reading and tax your eye movement; they signal rough going.

In order to succeed in the world of business, workers must learn to brush up on their networking skills. The network process has many benefits that you need to be aware of. These benefits range from finding a better job to accomplishing your job more easily and efficiently. Through networking you are able to expand the number of contacts who can help you. Networking means sharing news and opportunities. The Internet is the key to successful networking.

On the other hand, do not print a document with too short or extremely uneven lines.

In order to succeed in the world of
business, workers must learn
to brush up on their networking skills. The
network process has many
benefits you need to be
aware of.

Readers will suspect your ideas are incomplete, superficial, or even simple-minded.

4. Columns. Document text usually is organized in either single-column or multi-column formats. Memos, letters, and reports are usually formatted without columns, whereas documents that intersperse text and visuals (such as newsletters and magazines) work better in multicolumn formats.

Typography

Typeface
Readability of your text is crucial. Select a typeface, therefore, that ensures your text is

- legible
- attractive
- functional
- appropriate for your message
- complementary with accompanying graphics

The most familiar typefaces are the following four.

Times Roman	Frutiger
Helvetica	Palatino

Avoid using a typeface that looks like script, and don't mix and switch typefaces. The result makes your work look amateurish and disorganized (see Figure 6.1).

Type Size

Type size options are almost unlimited, depending on your software package and printer capabilities. Type size is measured in units called **points,** 72 points to the inch. The bigger the point size, the larger the type. Never print your letter or report in 6- or 8-point newspaper ad type or in a size larger than 12-point type. Here are some suggestions.

Times 8 point = footnotes/endnotes

Helvetica 12 point = letters, reports

Palatino 14 point = headings

Frutiger 22 point = title of your report

Type Styles

Type styles include boldface, italics, shadow, underlining, small caps, and shading.

Boldface
Italics
Shadow
Underlining
SMALL CAPS
Shading

Avoid overusing boldface and italics. Use them only when necessary and not for decoration. Do *not* underline the text unless absolutely necessary. Not only will too many special effects make your work harder to read, but you will lose the dramatic impact these features have to distinguish and emphasize key points that do deserve boldface or italic type.

Justification

Sometimes referred to as alignment, justification consists of left, right, full, and center options. Left-justified (also called **unjustified** or **ragged right**) is the preferred method because it allows space between words in lines of text to remain constant and is easier to read.

Our new website offers consumers a mall on the Internet. It gives shoppers access to our products and services and makes buying easy and fun. Our new website offers consumers a mall on the Internet. It gives shoppers access to our	Our new website offers consumers a mall on the Internet. It gives shoppers access to our products and services and makes buying easy and fun. Our new website offers consumers a mall on the Internet. It gives shoppers access to our
Left-justified text	**Right-justified text**

Heads and Subheads

Heads and subheads are brief descriptive words or phrases to introduce or summarize a document or a section or subsection within a document. They are typographical markers that signal starting points and major divisions in your document. Note how heads provide helpful landmarks for readers charting their course through a document, as in Figure 6.2. Without heads and subheads your work will look unorganized and cluttered. They should be grammatically parallel and not wordy. Note how these headings from a poorly organized proposal from the Acme Company are not parallel.

- What Is the Problem?
- Describing What Acme Can Do to Solve the Problem
- It's a Matter of Time . . .
- Fees Acme Will Charge
- When You Need to Pay
- Finding Out Who's Who

Revised, the heads are parallel and easier for a reader to understand and follow.

- A Brief History of the Problem
- A Description of Acme Solutions
- A Timetable Acme Will Follow
- A Breakdown of Acme's Fees
- A Payment Plan
- A Listing of Acme's Staff

Heads and subheads immediately attract attention and quickly inform readers about the function, scope, purpose, or contents of the document or section. The space around a heading is like an oasis for the reader, signaling both a rest and a new beginning. In designing a document with heads and subheads, follow these guidelines.

- Use a larger type size for heads than for text; major heads should be larger than subheads. If your text is in 10-point type, your heads can be in 16-point type and your subheads in 12- or 14-point type.

> ## Fourteen-Point Head
> ### Subhead in 12 Point
> Use larger type for heads than you do for text; major headings should be larger than subheads. If your text is in 10-point type, your heads may be in 14-point type and your subheads in 12.

- Modify type to differentiate sections. For example, for heads and subheads use uppercase, bold type, underlining, and changes in font style or type.
- Establish a horizontal position for a head, such as centered or aligned left, and keep it consistent throughout the document.
- Do not overuse heads (e.g., inserting them after each paragraph); having too many heads are as bad as having too few.

Lists

Placing items in a list helps readers by dividing, organizing, and ranking information. Lists emphasize important points and contribute to an easy-to-read page design. Lists can be numbered, lettered, or bulleted. Take a look at the memos, reports, and proposals in Chapters 3, 8, and 9 that effectively use lists.

Using Color

Using color in workplace documents is a good way to enhance readability, break up long segments of text, and tie important ideas together. Color can be effectively employed for borders and graphic accents, heading, titles, keywords, Internet addresses, sidebars, rules, and boxes that link related facts, figures, or information. Determine if using color serves a functional purpose or if it is merely a decoration. If it is window-dressing, stick with black and white.

Some Guidelines on Using Color

- Print a sample page because colors look different on a computer screen than they do on the printed page.
- Make sure text colors contrast sharply with background colors. Text colors should be dark, and background colors should be lighter.

- Use no more than two or three colors on a page, unless there are photographs, illustrations, or graphics that are multicolor.
- Avoid too many bright colors, which may be hard on the reader's eyes. Try to stick with "cool" colors, such as deep blue, turquoise, purple, forest green, and magenta. If you use bright colors such as bright reds, yellows, greens, and oranges, use them sparingly to call attention to important elements.
- Select colors that are culturally appropriate for international audiences (for more information, see page 204).

Three Rules of Effective Page Design: A Wrap Up

By following these three simple guidelines you can design effective documents for your readers.

1. Keep it simple. Don't try to impress your reader with visual effects—the overuse of italics or boldface (they will lose their purpose), fancy designs and scripts, printing everything in capitals or other large type size usually reserved for headlines.

2. Make it clear. Make your message logical and easy to understand by including heads, bulleted lists, numbered steps, and the like.

3. Have it flow. Help your readers get through and understand your document easily and, if necessary, go back and find information quickly. Don't cram too much information on a page (lots of visuals, including photos, several lists). Skimping on white space also makes your document difficult to digest. Steer clear of tight spacing and small type size.

The Purpose of Visuals

Now that you can design professional looking documents, learn how to integrate visuals for additional support. Here are several reasons why visuals can improve your work; each point is graphically reinforced in Figure 6.3 (p. 179).

1. Visuals arouse readers' immediate interest. They catch the reader's eye quickly by setting important information apart and by giving relief from sentences and paragraphs. Note the eye-catching quality of the visual in Figure 6.3.

2. Visuals increase readers' understanding by simplifying concepts. A visual *shows* ideas whereas a verbal description only *tells* about them in the abstract. Visuals are especially important and helpful if you must explain a technical process to a non-specialist audience. Visuals help readers see percentages, trends, comparisons, and contrasts. Figure 6.3, for example, shows at a glance the growth of e-commerce.

3. Visuals are especially important for non-native speakers of English and multi-cultural audiences. Given the international audience for many business documents, visuals will make your communication with them easier and clearer. (See Terri Smith Ruckel's report in Figure 9.1, pp. 291–308.)

A line-and-bar chart depicting the growth of e-commerce compared to traditional businesses.   **Figure 6.3**

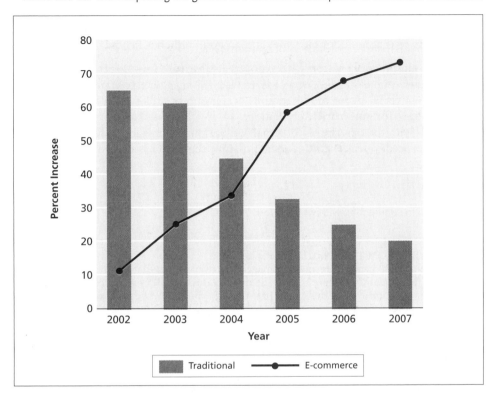

4. Visuals emphasize key relationships. Through their arrangement and form, visuals quickly show contrasts, similarities, growth rates, and downward and upward movements, as well as fluctuations in time, money, and space.

5. Visuals condense and summarize a large quantity of information into a relatively small space. A visual also allows you to streamline your message by saving words and time. It can record data in far less space than it would take to describe those facts in words alone. Note how in Figure 6.3 the growth rates of two different types of businesses are expressed and documented.

6. Visuals are highly persuasive. Visuals have sales appeal. They can convince readers to buy your product or service or to accept your point of view and reject your competitor's.

Choosing Effective Visuals: Some Precautions

Select your visuals carefully. Various computer software programs allow you to select, create, and introduce visuals. The following suggestions will help you to choose effective visuals.

1. Use visuals only when they are relevant for your purpose and audience. Never include a visual simply as a decoration. A short report on fire drills, for example, does not need a picture of a fire station. Avoid any visual that is too technical for your readers or that includes more detail than your audience needs.

2. Use visuals in conjunction with—not as a substitute for—written work. Visuals do not always take the place of words. In fact, you may need to explain information contained in a visual. A set of illustrations or a group of tables alone may not satisfy readers looking for summaries, evaluations, or conclusions. Note how the visual in conjunction with the description of a magnetic resonance imager (MRI) in Figure 6.4 makes the procedure easier to understand than if the writer had used only words or

Figure 6.4   A visual used in conjunction with written work.

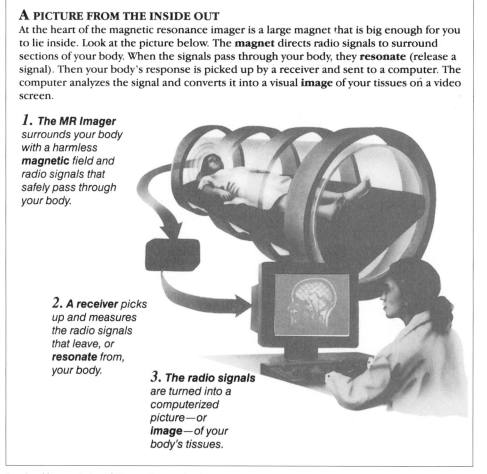

A PICTURE FROM THE INSIDE OUT
At the heart of the magnetic resonance imager is a large magnet that is big enough for you to lie inside. Look at the picture below. The **magnet** directs radio signals to surround sections of your body. When the signals pass through your body, they **resonate** (release a signal). Then your body's response is picked up by a receiver and sent to a computer. The computer analyzes the signal and converts it into a visual **image** of your tissues on a video screen.

1. The MR Imager surrounds your body with a harmless *magnetic* field and radio signals that safely pass through your body.

2. A receiver picks up and measures the radio signals that leave, or *resonate* from, your body.

3. The radio signals are turned into a computerized picture—or *image*—of your body's tissues.

Reprinted by permission of Krames Communications.

only a visual. This visual and verbal description is appropriately included in a brochure teaching patients about MRI procedures.

3. Experiment with several visuals. Evaluate a variety of options before you select a particular visual. For instance, a graphics software package such as PowerPoint (see Chapter 10, pp. 320–322) will allow you to represent statistical data in a number of ways. Preview a few different versions of a visual or even different types of visuals to determine which one would be best. Test your visual first to identify parts and find the proper scale. Never distort a visual for emphasis or decoration or choose one that contradicts your written work.

4. Be prepared to revise and edit your visuals. Just as you draft, revise, and edit your written work to meet your audience's needs, create several versions of your visual to get it right. Expect to change shapes or proportions; experiment with different colors, shadings, labels, and sizes. Avoid any visuals that may distract the reader or contradict your message. Make sure each visual is accurate and ethical, not unclear, distorted, or exaggerated. (See "Using Visuals Ethically" on pages 198–202.)

5. Always use high-quality visuals. Your visuals should be clear, easy-to-read, attractive, and relevant. If you photocopy, download, or scan a visual, make sure the copy is clear and readable and does not cut off any part of the original.

6. Consider how your visuals will look on the page. Don't cram visuals onto a page or allow them to spill over your text or margins.

Inserting and Writing About Visuals: Some Guidelines

The following guidelines will help you to (1) identify, (2) insert, (3) introduce, and (4) interpret visuals for your readers. By observing these guidelines, you can use visuals more effectively and efficiently.

Identify Visuals

Give each visual a number and caption (title) that indicates the subject or explains what the visual illustrates. An unidentified visual is meaningless. A caption helps your audience to interpret your visual—to see it with your purpose in mind.

- Use a different typeface (bold) and size in your caption than what you use in the visual itself.
- Include key words about the function and the subject of your visual in a caption.
- If you use a visual that is not your own work, give credit to your source (newspaper, magazine, textbook, company, federal agency, individual, or website). If your paper or report is intended for publication, you must obtain permission to reproduce copyrighted visuals from the copyright holder.

Below are some examples of figure numbers and captions.

- Figure 2. Paul Jordan's work schedule, January 15–23
- Figure 4.6. The proper way to apply for a small business loan
- Figure 12. Income Estimation Figures for North Point Technologies

Insert Visuals Appropriately

Here are some guidelines to help you incorporate the visuals in the most appropriate places for your readers.

- Never introduce a visual *before* a discussion of it; readers will wonder why it is there. Use a sentence or two to introduce your visuals.
- Always mention in the text of your paper or report that you are including a visual. Tell readers where it is found—"below," "to the right," "at the bottom of the page."
- Place visuals as close as possible to the first mention of them in the text. Try not to put a visual more than one page after the discussion of it. Don't wait two or three pages to present it.
- Use an appropriate size for your visual. Don't make it too large or too small.
- Center your visual and, if necessary, box it. Leave at least 1 inch of white space around it. Squeezing visuals toward the left or right margins looks unprofessional.
- Never collect all your visuals and put them in an appendix. Readers need to see them at those points in your discussion where they are most pertinent.

Introduce Your Visuals

Refer to each visual by its number and, if necessary, mention the title as well. In introducing the visual, though, do not just insert a reference to it, such as "See Figure 3.4" or "Look at Table 1." Help readers to understand the relationships in your visual. Here are two ways of writing a lead-in sentence for a visual.

Poor: Our store saw a dramatic rise in the shipment of electric ranges over the five-year period as opposed to the less impressive increase in washing machines. (See Figure 3.)

The visual just trails insignificantly behind.

Better: As Figure 3 shows, our store saw a dramatic rise in the shipment of electric ranges over the five-year period as opposed to the less impressive increase in washing machines.

Mentioning the visual in Figure 3 alerts readers to its presence and function in your work and helps them to better understand it.

Interpret Your Visuals

Give readers help in understanding your visual and in knowing what to look for. Let them know what is most significant about the visual. Do not expect the visual to explain itself. Inform readers what the numbers or images in your visual mean. What types of conclusions can you and your readers logically make after seeing the visual?

In a report on the benefits of vanpooling, the writer supplied the following visual, a table:

TABLE 1.2   Travel Time (in minutes): Automobile Versus Vanpool

Private Automobile	Vanpool
25	32.5
30	39.0
35	45.5
40	52.0
45	58.5
50	65.0
55	71.5
60	78.0

Source: U.S. Department of Transportation. *Increased Transportation Efficiency Through Ridesharing: The Brokerage Approach* (Washington, D.C., DOT-OS—40096): 45.

Then, explaining the table, the writer called attention to it in the context of the report on transportation efficiency.

Although, as Table 1.2 above suggests, the travel time in a vanpool may be as much as 30 percent longer than in a private automobile (to allow for pickups), the total trip time for the vanpool user can be about the same as with a private automobile because vanpools eliminate the need to search for parking spaces and to walk to the employment site entrance.[1]

Two Categories of Visuals

Visuals can be divided into two categories—tables and figures. **Tables** arrange information—numbers and/or words—in parallel columns or rows for easy comparison of data. Anything that is not a table is considered a figure. **Figures** include graphs, circle charts, bar charts, organizational charts, flow charts, pictographs, maps, photographs, and drawings. Expect to use both in your work.

Tables

Tables are parallel columns or rows of information organized and arranged into categories to show changes in time, distance, cost, employment, or some other distinguishable or quantifiable variable. They allow readers to compare a great deal of information in a compact space. Tables also summarize material for easy recall—causes of wars; provisions of a law; or differences between a common cold, the flu, and pneumonia.

---

[1]James A. Devine, "Vanpooling: A New Economic Tool," *AIDC Journal*.

Parts of a Table

To use a table properly, you need to know the parts that constitute it. Refer to Table 6.1, which labels these parts, as you read the following:

- The main **column** is "Amount Needed to Satisfy Minimum Daily Requirement," and the **subcolumns** are the protein sources for which the table gives data.
- The **stub** refers to the first vertical column on the left side. The stub column heading is "Source." The stub lists the foods for which information is broken down in the subcolumns.
- A **rule** (or line) across the top of the table separates the headings from the body of the table.

Guidelines for Using Tables

When you include a table in your work, follow these guidelines.

- Number the tables according to the order in which they are discussed in the text (Table 1, Table 2, Table 3). Tables are numbered separately, from figures (charts, graphs, photos) in your text.
- Include the table on the same page, where it is most appropriate, whenever possible.
- Keep your table on one page; it is difficult for readers to follow a table spread across different pages.
- Give each table a concise and descriptive title to show exactly what is being represented or compared.
- Use words in the **stub** (a list of items about which information is given), but put numbers under column headings. The "Source" column in Table 6.1 is the stub.

TABLE 6.1   Parts of a Table

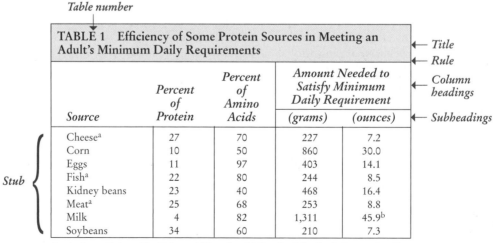

Table number

TABLE 1 Efficiency of Some Protein Sources in Meeting an Adult's Minimum Daily Requirements				
Source	*Percent of Protein*	*Percent of Amino Acids*	*Amount Needed to Satisfy Minimum Daily Requirement*	
			(grams)	*(ounces)*
Cheese[a]	27	70	227	7.2
Corn	10	50	860	30.0
Eggs	11	97	403	14.1
Fish[a]	22	80	244	8.5
Kidney beans	23	40	468	16.4
Meat[a]	25	68	253	8.8
Milk	4	82	1,311	45.9[b]
Soybeans	34	60	210	7.3

Labels: *Title*, *Rule*, *Column headings*, *Subheadings*, *Stub*, *Origin of data*, *Footnotes*

Source: From *Biology: The Unity and Diversity of Life,* 4th edition by C. Starr and R. Taggart. Copyright © 1987. Reprinted with permission of Brooks/Cole, a division of Thomson Learning: **www.thomsonrights.com**. Fax 800-730-2215.

[a] = Average value
[b] = Equivalent of 6 cups

- Supply footnotes (often indicated by small raised letters: [a,b]) if something in the table needs to be qualified—for example, the number of cups of milk in Table 6.1. Then put that information below the table.
- Arrange the data you want to compare vertically; it is easier to read down than across a series of rows.
- Place tables at the top (preferable) or bottom of the page and center them on the page rather than placing them up against the right or left margin.
- Don't use more than five or six columns; tables wider than that are more difficult for readers to use.
- Round off numbers in your columns to the nearest whole number to assist readers in following and retaining information.
- Always give credit to the source (the supplier of the statistical information) on which your table is based.

Figures

As mentioned before, any visual that is not a table is classified as a **figure.** The types of figures we examine here are

- line graphs
- circle, or pie, charts
- bar charts
- organizational charts
- flow charts
- pictographs
- photographs
- drawings

Line Graphs

Graphs transform numbers into pictures. They take statistical data presented in tables and put them into rising and falling lines, steep or gentle curves.

Functions of Line Graphs
Graphs vividly portray information that changes, such as

- costs
- sales
- fluctuations
- profits
- distributions
- increases and decreases in (e.g., jobs, houses, etc.)
- employment
- energy levels
- temperatures

Simple Line Graphs
Basically, a simple graph consists of two sides—a **vertical axis** and a **horizontal axis**—that intersect to form a right angle, as in Figure 6.5. The space between the

Figure 6.5   A simple line graph showing the amount of snowfall in Springfield from November 2008 to April 2009.

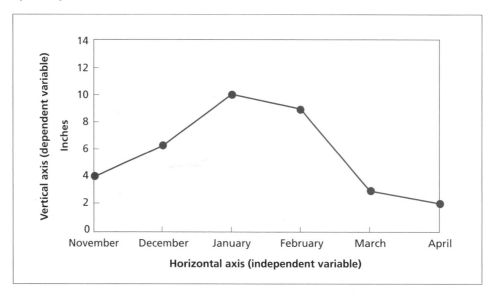

two axes contains the picture made by the graph—the amount of snowfall in Springfield between November 2008 and April 2009. The vertical line represents the **dependent variable** (the snowfall in inches), the horizontal line, the **independent variable** (time in months). The dependent variable is influenced most directly by the independent variable, which almost always is expressed in terms of time or distance. The vertical axis is read from bottom to top; the horizontal axis from left to right.

Multiple-Line Graphs

The graph in Figure 6.5 contains only one line per category. But a graph can have multiple lines to show how a number of dependent variables (conditions, products) compare with each other.

The six-month sales figures for three salespeople can be seen in the graph in Figure 6.6 (p. 187). The graph contains a separate line for each of the three salespersons. At a glance readers can see how the three compare and how many dollars each generated per month. Note how the line representing each salesperson is clearly differentiated from the others by symbols and colors. Each line is clearly tied to a **legend** (an explanatory key below the graph) specifying the three salespersons.

Guidelines for Creating a Graph

1. Use no more than three lines in a multiple-line graph so that readers can interpret the graph more easily. If the lines run close together, use a legend to identify individual lines.
2. Label each line to identify what it represents.
3. Keep each line distinct in a multiple-line graph by using different colors, dots or dashes, or symbols. Note the different symbols in Figure 6.6.

A multiple-line graph showing sales figures for the first six months of 2008 for three employees.   **Figure 6.6**

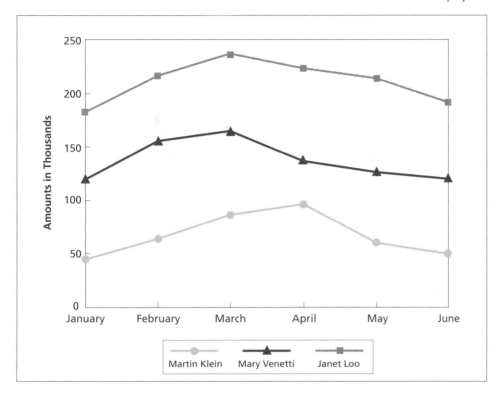

4. Make your graph data points large enough to show a reasonable and ethical number of plotted points (using only three or four data points may distort the evidence).
5. Keep the scale consistent and realistic. If you start with hours, do not switch to days or vice versa. If you are recording annual rates or accounts, do not skip a year or two in the hope that you will save time or be more concise. Do not use, for example, 1999, 2001, 2002, 2003, 2006, 2009. Include all the years you are surveying or equal multiples of them (such as 2000, 2002, 2004, 2006, 2008).

Charts

Among the most frequently used charts are (1) circle, or pie, charts, (2) bar charts, (3) organizational charts, and (4) flow charts.

Circle Charts

Circle charts are also known as **pie charts**, a name that descriptively points to their construction and interpretation. Tables are more technical and detailed than circle charts. Figure 6.7 shows an example of a pie chart used in a government document. A table or graph with a more detailed breakdown of, say, a city's budget would be much more appropriate for a technical audience (auditors, budget and city planners).

Figure 6.7   A three-dimensional circle chart showing the breakdown by department of the proposed Midtown City Budget for 2008.

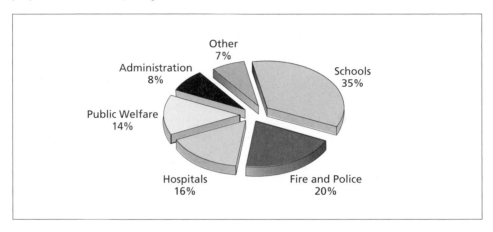

The full circle, or pie, represents the whole amount (100 percent or 360 degrees) of something: the entire budget of a company or a family, a population group, an area of land, the resources of an organization or institution. Each slice or wedge represents a percentage or portion of the whole.

A circle chart effectively allows readers to see two things at once: the relationship of the parts to one another and the relationship of the parts to the whole.

Preparing a Circle Chart

Follow these seven rules to create and present your circle chart.

1. Keep your circle chart simple.   Don't try to illustrate technical statistical data in a pie chart. Pie charts are primarily used for general audiences.

2. Do not divide a circle, or pie, into too few or too many slices.   If you have only three wedges, use another visual to display them (a bar chart, for example, discussed below). If you have more than seven or eight wedges, you will divide the pie too narrowly, and overcrowding will destroy the dramatic effect. Instead, combine several slices of small percentages (2 percent, 3 percent, 4 percent) into one slice labeled "Other," "Miscellaneous," or "Related Items."

3. Make sure the individual slices total 100 percent, or 360 degrees.   The breakdown percentages to represent a family's budget might be as follows:

Category	Percentage	Angle of slice
Housing	25%	90.0°
Food	22%	79.2°
Energy	20%	72.0°
Clothes	13%	46.8°
Health Care	12%	43.2°
Miscellaneous	8%	28.8°
Total	100%	360°

4. Put the largest slice first, at the 12 o'clock position, then move clockwise with proportionately smaller slices.   Schools occupy the largest slice in Figure 6.7 because they receive the biggest share of taxes.

5. Label each slice of the pie horizontally.   Do not put in a label upside down or slide it in vertically. If the individual slice of the pie is small, draw a connecting line from the slice to a label positioned outside the pie.

6. Shade, color, or cross-hatch slices of the pie to further separate and distinguish the parts.   Figure 6.7 effectively uses color. But be careful not to obscure labels and percentages; also make certain that adjacent slices can be distinguished readily from each other. Do not use the same color or similar colors for two slices.

7. Give percentages for each slice to further assist readers, as in Figure 6.7.

Bar Charts

A bar chart consists of a series of vertical or horizontal bars that indicate comparisons of statistical data. For instance, in Figure 6.8 vertical bars depict increases in number of working mothers. Figure 6.9 uses horizontal bars to depict the nation's top 20 metropolitan areas, based on building permits. The length of the bars is determined according to a scale that your computer software can easily calculate.

Vertical bar chart.   **Figure 6.8**

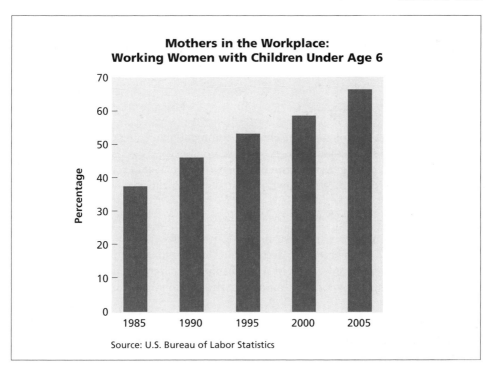

Figure 6.9   Horizontal bar chart.

Organizational Charts

An organizational chart pictures the chain of command in a company or agency, with the lines of authority stretching down from the chief executive, manager, or administrator to assistant manager, department heads, or supervisors to the work force of employees. Figure 6.10 (p. 191) shows a hospital's organizational chart for its nursing services.

Organizational charts have many advantages:

- informing employees and customers about the makeup of a company
- depicting the various offices, departments, and units
- showing where people work in relationship to each other in a business
- coordinating employee efforts in routing information to appropriate departments

Flow Charts

A **flow chart** displays the stages in which something is manufactured, or is accomplished, develops, or operates. Flow charts are highly effective in showing the steps of a procedure. They can also be used to plan the day's or week's activities.

An organizational chart representing critical care nursing services at Union General Hospital.   **Figure 6.10**

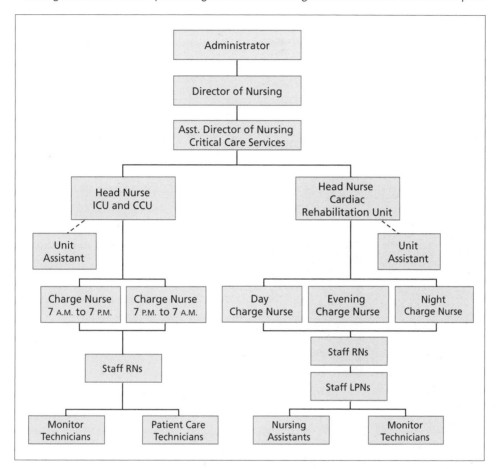

A flow chart tells a story with arrows, boxes, and sometimes pictures. Boxes are connected by arrows to show the stages of a process. Flow charts often proceed from left to right and back again, as in the one below, showing the steps to be taken before graduation.

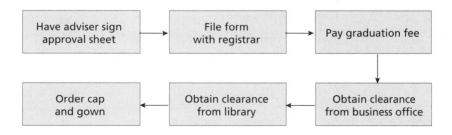

Flow charts can also be constructed to read from top to bottom. Computer programming instructions often are written that way. See, for example, Figure 6.11,

Figure 6.11   A programming flow chart showing steps in writing a research paper.

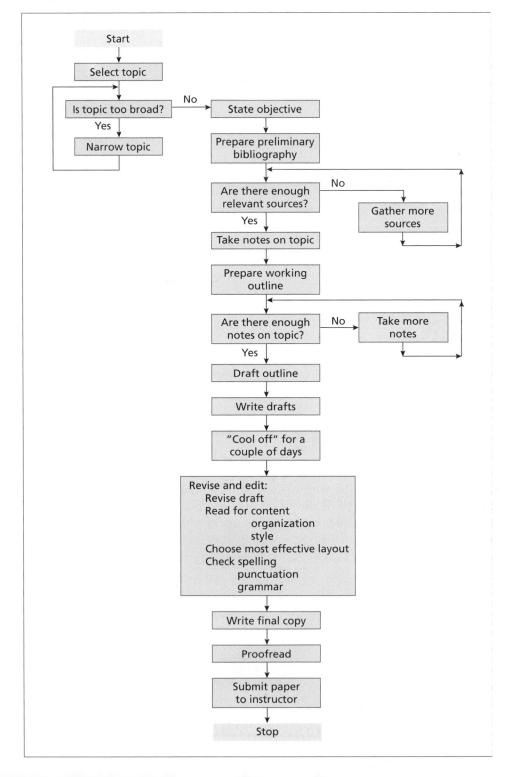

A pictograph showing the growth of one state's retirement assets.   **Figure 6.12**

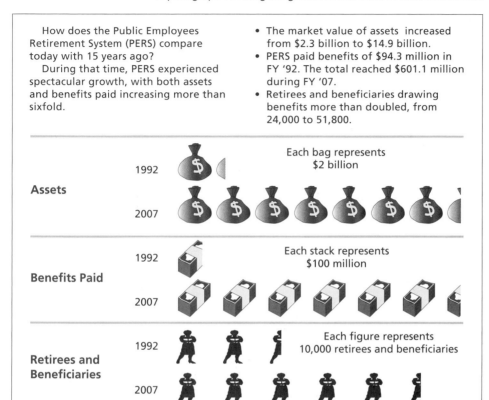

How does the Public Employees Retirement System (PERS) compare today with 15 years ago?
  During that time, PERS experienced spectacular growth, with both assets and benefits paid increasing more than sixfold.

- The market value of assets increased from $2.3 billion to $14.9 billion.
- PERS paid benefits of $94.3 million in FY '92. The total reached $601.1 million during FY '07.
- Retirees and beneficiaries drawing benefits more than doubled, from 24,000 to 51,800.

Assets
1992
2007
Each bag represents $2 billion

Benefits Paid
1992
2007
Each stack represents $100 million

Retirees and Beneficiaries
1992
2007
Each figure represents 10,000 retirees and beneficiaries

Source: PERS Member Newsletter. Official Publication of the Public Employees' Retirement System of Mississippi. By permission of Public Employees' Retirement System.

which uses a computer program's flow chart feature to show the steps a student must follow in writing a research paper.

Pictographs

A **pictograph** uses picture symbols (called **pictograms**) to represent differences in statistical data, as in Figure 6.12, above. A pictograph repeats the same symbol or icon to depict the quantity of items being measured. Each symbol stands for a specific number, quantity, or value. Pictographs are visually appealing and dramatic and are far more appropriate for a nontechnical than a technical audience.

When you create a pictograph, follow these three guidelines:

1. Choose an appropriate symbol for the topic—such as handheld PCs to represent the increase in the number of sales of BlackBerries.
2. Always indicate the precise quantities involved by placing numbers after the pictures or at the top of the visual.

3. Increase the number of symbols rather than their sizes because differences in size are often difficult to construct accurately and harder for readers to interpret.

Photographs

Correctly taken or scanned, photographs are an extremely helpful addition to job-related writing. A photograph's chief virtues are realism and clarity. A photo can

- show what an object looks like
- demonstrate how to perform a certain procedure
- compare relative sizes and shapes of objects
- compare and contrast scenes or procedures

Note how the photograph in Figure 6.13, below, does the first three.

Digital cameras allow you to supply professional-looking, customized photos with your written work. As with other cameras, you point and shoot with a digital camera. But unlike a film camera, the digital camera lets you review every photo you take before you print it. You can shoot a photo, take it in slow motion, and play it back, which allows you to scroll, zoom in, blow it up, adjust darkness or light, even change exposures.

Figure 6.13   A photo showing how to perform a procedure.

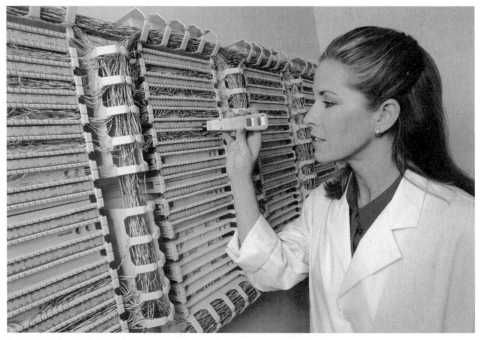

Source: Corbis Royalty Free *(http://www.corbis.com).*

Make sure when using a digital camera or camera phone that it is set to the correct photo-quality level. The poorer the quality, the more pictures your camera will hold. However, if you use photographs for a newsletter or company magazine or want to print photographic-quality images, you will have to set your camera on a higher quality level.

Before you take any photo, though, decide how much foreground and background image you need. Include only the details that are *necessary* and *relevant* for your purpose. Observe the following guidelines:

1. Focus your camera on the most important part of the image.
2. Select the correct angle.
3. Give the right amount of detail. Pictures that include clutter compete for the reader's attention and detract from the subject.
4. Take the picture from the right distance. If you need a shot of a three-story office building, your picture may show only one or two stories if you are standing too close to the building when you photograph it. Standing too far away from an object, however, means that the photograph will reduce the object's importance and record unnecessary details.

Drawings

Drawings can show where an object is located, how a tool or machine is put together, or what signals are given or steps taken in a particular situation. A drawing can be simple, like the one in Figure 6.14 (p. 196), which shows readers exactly where to place smoke detectors in a house.

A more detailed drawing can reveal the interior of an object. Such sketches are called **cutaway drawings** because they show internal parts normally concealed from view. Figure 6.15 (p. 197) is a cutaway drawing of an antilock brake system.

Another kind of sketch is an **exploded drawing,** which blows the entire object up and apart to show how the individual parts are arranged. An exploded drawing comes with most computers and uses **callouts,** or labels, to identify the components. The labels are often attached to the drawing with arrows or lines. As the name suggests, the labels "call out" the parts so that readers can identify them quickly.

Guidelines for Using a Drawing

1. Keep your drawing simple. Include only as much detail as your reader will need to understand what to do, be it to assemble or to operate a mechanism. Do not include any extra details.
2. Clearly label all parts so that your reader can identify and separate them.
3. Decide on the most appropriate view of the object you want to illustrate—aerial, frontal, lateral, reverse, exterior, interior—and indicate in your title which view it is.
4. Keep the parts of the drawing proportionate unless you are purposely enlarging one section.

Figure 6.14   A simple drawing showing where to place smoke detectors in a house.

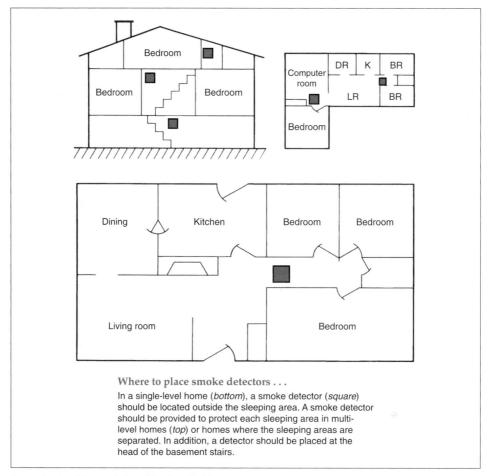

Where to place smoke detectors . . .

In a single-level home (*bottom*), a smoke detector (*square*) should be located outside the sleeping area. A smoke detector should be provided to protect each sleeping area in multi-level homes (*top*) or homes where the sleeping areas are separated. In addition, a detector should be placed at the head of the basement stairs.

Source: Reprinted by permission of *Southern Building.*

Clip Art

Clip art (or icons) refers to ready-to-use electronic images. These small cartoon-style representations and photographs, such as the ones shown in Figure 6.16 (p. 198) depict almost anything you can think of under such headings as energy, government, leisure, health, money, the outdoors, food, technology, transportation. A free clip art and photo-illustration database can be found at Yahoo Picture Gallery at *http://gallery .yahoo.com.*

When you use clip art, follow these guidelines:

1. **Choose simple, easy-to-understand icons.** Select an image that conveys your idea quickly and directly. Avoid using an icon of an unfamiliar object or of a drawing or silhouette that might confuse your audience.
2. **Use clip art functionally.** Do not insert clip art as decorations. Including too many will make your work look unprofessional.

Cutaway drawing of an antilock brake system.   **Figure 6.15**

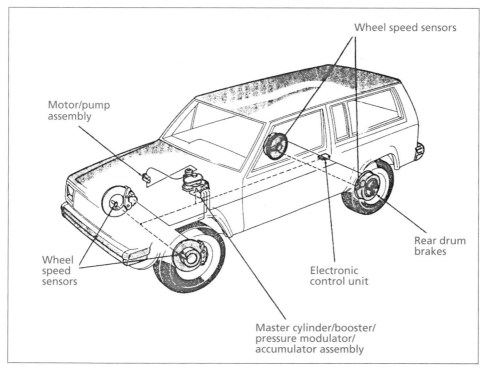

Wheel speed sensors

Motor/pump
assembly

Rear drum
brakes

Wheel
speed
sensors

Electronic
control unit

Master cylinder/booster/
pressure modulator/
accumulator assembly

Source: Chilton Automotive Books. *Jeep Wagoneer/Comanche 1984–1996 Repair Manual.* Chilton is a registered trademark of Cahners Business Information, a Division of Reed Elsevier Inc. and has been licensed to W. G. Nichols, Inc. Reprinted with permission.

3. **Make sure the clip art is relevant for your audience and your message.** A clip art airplane does not belong in a technical report on fuel capacity.
4. **Keep your clip art professional.** Some clip art is humorous, even silly, which may not be appropriate for a professional business report or proposal.

Using Visuals Ethically

Make sure your visuals, whether you create or import them, are ethical. Ethical visuals convey and interpret statistical information and other types of data, products and equipment, and locations, without misinterpretation. Your visuals should neither distort events and data nor mislead readers. Ethical visuals should be:

- accurate
- honest, fair
- complete
- appropriate
- easy to read
- clearly labeled
- uncluttered
- consistent with conventions

To ensure that your visuals are honest, accurate, and easy to read, observe the following guidelines.

Figure 6.16   Examples of clip art.

Photos

- Do not distort a photo by omitting key details or by misrepresenting dimensions, angles, sizes, or surroundings or by superimposing one image over another.
- Do not "doctor" a photo or show an off-site location, studio, or lab and then claim it as an "actual" location.
- Never take a photo of an individual for business purposes (e.g., putting it in a newsletter, ad, on a website) without his or her permission.
- Don't counterfeit or subtly alter a company's logo to sell, distribute, or promote an imitation as the real thing.

Graphs

- Don't distort a graph by plotting it in misleading or unequal intervals. For example, omitting certain years or dates to hide a decline in profits (contrast Figure 6.17, and its misleading interpretation, with the ethical revision in Figure 6.18).
- Include information in correct chronological sequence along the horizontal axis.
- Don't switch the type of information usually given along the vertical axis with the horizontal axis.
- Don't project growth or increases without having reliable and valid reasons.

Bar Charts

- Make sure the height and width of each bar truthfully represents the data it purports to. That is, don't make one of the bars larger to maximize the profits, products, or sales in any one year.
- Show bars for every year (or other sales period) covered. Note how the unethical bar chart (and accompanying text) in Figure 6.19 violates the rule but the chart in Figure 6.20 faithfully represents the data.

Unethical visual and misleading interpretation.   **Figure 6.17**

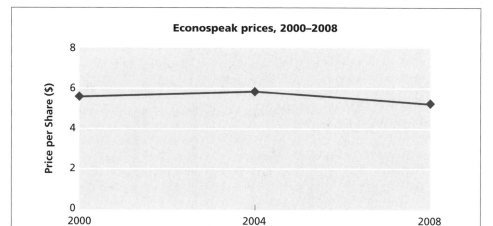

Econospeak's prices during 2000–2008 have been stable, resting securely at about $5.60. The graph above illustrates the stability of Econospeak's stock. Given our steady market, we believe shareholders will be confident in our recent decision to proceed with Econospeak's expansion into international markets.

Ethical revision of Figure 6.17.   **Figure 6.18**

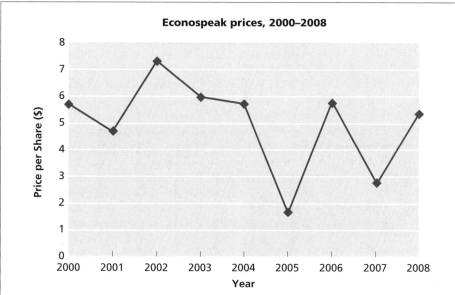

Econospeak's prices have not been as stable over 2000–2008 as we would have liked. The graph above illustrates the difficulties the company has faced in the market in the past decade, resulting in shifts by shareholders. We believe, however, that Econospeak's expansion into the global market will be possible by 2009.

Figure 6.19   Unethical visual.

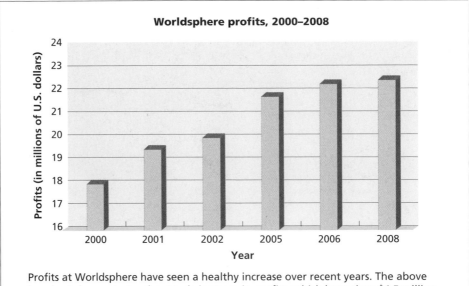

Profits at Worldsphere have seen a healthy increase over recent years. The above bar chart demonstrates the steady increase in profits, which have risen $4.5 million since 1998. Given the profit history of Worldsphere, our investors can be confident of future profits and the security of their stock with our company.

Figure 6.20   Ethical revision of Figure 6.19.

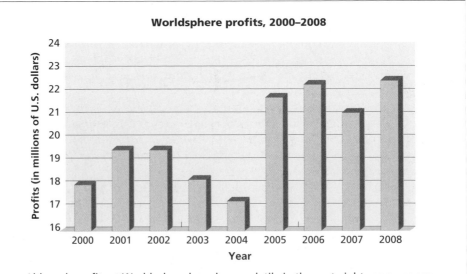

Although profits at Worldsphere have been volatile in the past eight years, we are now at our highest profit margin yet. The bar chart above shows the effects of market difficulties for the period 2000–2008, when the industry suffered major cutbacks. However, Worldsphere achieved a successful turnaround in 2003, with profits regaining strength due to revised marketing.

Pie Charts

- Don't use 3-D to distort the thickness and unjustified emphasis of one slice of the circle to misleadingly deemphasize others.
- Don't conceal negative information (losses, expenses, etc.) by silently including the information in another category or slice or lumping it into a category marked "other" or "miscellaneous."
- Make sure percentages match the number and size of the wedges or slices of the pie chart. Note how a larger expense for Guest Speakers (35% of budget) is unethically misrepresented in Figure 6.21 by using a smaller-sized wedge, while the expenses for Venue Rental are actually less than for Guest Speaker expenses but drawn larger to misrepresent costs. Figure 6.22 is an ethical revision of Figure 6.21.

Drawings

- Avoid any clutter that hides features.
- Label all parts correctly.
- Do not omit or shadow any necessary parts.
- Draw an object accurately. Tell readers if your drawing is the actual size of the object or equipment or if it is drawn to scale. Provide a scale.

Using Appropriate Visuals for International Audiences

Choose your visuals for international readers just as carefully as you do your words. Visuals should be designed to help international audiences understand your message clearly and without bias. Clear, effective visuals are crucial to your success in the global marketplace. See page 13 of Terri Ruckel's report in Chapter 9 to learn more about communicating effectively with graphics for an international audience in the U.S. work force.

Keep in mind that visuals do not automatically transfer from one culture to another. They are not boilerplate images that have the same meaning for all readers across the globe. Like words, visuals—photos, clip art, designs, colors, and other graphic devices—may have one meaning or use in the United States and a radically different one in other countries.

Visuals reflect the national, ethnic, and racial traditions of a particular country. Since they are symbols of cultural identity and value, make sure you respect your readers' visual communication. Investigate the meaning and cultural significance of a visual before you include it. Consult a native speaker from your audience's country or foreign language instructor to see if your visuals are culturally acceptable.

To communicate appropriately and respectfully with international readers through visuals and other graphics, follow these guidelines:

1. Do not use any images that ethnically or racially stereotype your readers. A large clothing manufacturer was criticized by the Asian American community when it released T-shirts depicting two men with rice paddy hats as laundromat owners. Depict-

Chapter 6 Additional Activities, located at **college.cengage .com/pic/ kolinconcise2e**, include a global-focused activity titled "Designing Visuals for International Audiences."

Figure 6.21   Unethical visual of pie chart.

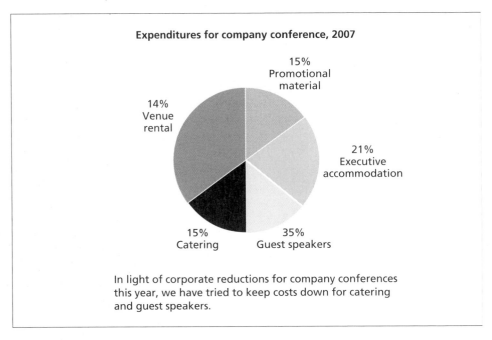

Expenditures for company conference, 2007

In light of corporate reductions for company conferences this year, we have tried to keep costs down for catering and guest speakers.

Figure 6.22   Ethical revision of Figure 6.21.

Expenditures for company conference, 2007

Though they took the largest share of company expenditures for the conference this year, the fees we paid for guest speakers may net our greatest advantages. These speakers have shown us ways to maximize our profits in expanding markets.

ing Native Americans through clip art images of red-faced chiefs is also insulting. Rather than using an ethnic pictogram, use neutral stick figures or appropriate clip art.

2. Be respectful of religious symbols and images. A cross to a U.S. audience represents a church, first aid, or a hospital, and, when the cross is a large red one on a white background, the Red Cross. But the symbol for that humanitarian agency in Turkey and in other Muslim countries is the crescent. Portraying a smiling Buddha to sell products is considered disrespectful to residents in Southeast Asia.

3. Avoid using culturally insensitive or objectionable photographs. What is acceptable in photographs in the United States may be taboo elsewhere in the world. Portraying men and women eating together at a business conference is unacceptable in Saudi Arabia. Moreover, depicting individuals seated with one leg crossed over the other is regarded as disrespectful in many countries in the world.

4. Be cautious about importing images/photos with hand gestures in your work, especially in manuals or other instructional materials. Many gestures are culture-specific; they do not necessarily mean the same thing in other countries that they do in the United States. Table 6.2 lists cultural differences around the globe for some common gestures.

5. Avoid any icons or clip art that international readers would misunderstand. Animal symbolism varies extensively from one country to another. In the United States, an owl can stand for wisdom, thrift, and memory, while in Japan, Romania, and some African countries it is a symbol for death. A software program showing the icon of a mailbox (below)

for the designation "You Have Mail" confused readers in some countries who thought the icon represented a birdhouse. A better alternative would be an icon of an envelope.

6. Be careful using directional signs and shapes. While road signs tend to be fairly recognizable throughout the world—for example, the octagon is the shape for the stop sign—there are country-specific signs and code books. For instance, a pennant on U.S. highways, exclusively signaling a **no-passing zone**, may not

TABLE 6.2   Different Cultural Meanings of Various Hand Gestures

Gesture	Meaning in the United States	Meaning in Other Countries
OK sign Index finger joined to thumb in a circle	All right; agreement	Sexual insult in Brazil, China, Germany, Russia; sign for zero, worthlessness in France
Thumbs-up	A winning gesture; good job; approval	Offensive gesture in Muslim countries
Pointing with index finger	This way; pay attention; turn the page	Rude, insulting in Japan; in Saudi Arabia used only to refer to animals, not people

have the same meaning in Nigeria or India. The symbol below points to a railroad crossing for American readers but would baffle an audience in the Czech Republic.

7. Don't offend international readers by using colors that are culturally inappropriate. A color may have a far different meaning for one group than it does for another. For instance, emphasizing the color red in a document aimed at Native Americans or the color yellow in a document for Asians or Asian Americans is likely to be seen as offensive. So, too, overusing red in a document for Russian or Eastern European readers may be interpreted as a deliberate reference to Communism.

Always do some research (on the Internet or by consulting with a representative from the audience you want to reach) to find out if the color scheme you have chosen for your document is appropriate for your target audience. For example, purple is the color of death and mourning in Thailand, as is white in China. Although orange is the symbolic color of Northern Ireland, you should avoid it when writing to readers outside of Northern Ireland.

8. Avoid confusing an international audience with punctuation and other writing symbols used in the United States. Punctuation use and symbols vary widely around the globe. Elipses (**. . .**) and slashes **/** (and/or) are punctuation symbols that may not be a part of the language your international audience reads and writes. Similarly, be careful about using the following graphic symbols, familiar to writers and speakers of U.S. English, but not necessarily to an international audience:

#	pound	&	ampersand (and)
©	copyright	*	asterisk
™	trademark	%	percentage

Include a glossary or a key to these symbols or any other graphics your reader may not understand or revise your sentences to avoid these symbols.

As this chapter has emphasized, a major part of your work-related writing requires you to select an effective layout and design as well as appropriate visuals for both your U.S. and international readers.

 Revision Checklist

☐ Arranged information in the most logical, easy-to-grasp order.
☐ Left adequate, eye-pleasing white space in text and margins to frame document.
☐ Maintained pleasing, easy-to-read line length and spacing.
☐ Chose appropriate typeface for message and document.
☐ Did not mix typefaces.
☐ Used effective type size, neither too small nor too large.
☐ Inserted heads and subheads to organize information for reader.
☐ Supplied lists, bullets, numbers to divide information.
☐ Chose colors carefully and with a specific purpose.
☐ Selected most effective visual (table, chart, graph, drawing, photograph) to represent information the audience needs.
☐ Drafted and edited visual until it meets readers' needs.
☐ Made sure every visual is attractive, clear, complete, and relevant.
☐ Gave each visual a number, a title, and, where necessary, a legend.
☐ Inserted visual near the written description or commentary to which it pertains.
☐ Introduced and interpreted each visual in appropriate place in report or paper.
☐ Acknowledged sources for visuals.
☐ Used and interpreted visuals ethically.
☐ Selected visuals and colors that respect the cultural traditions of international readers.
☐ Avoided using any images of hand gestures that may be offensive to an international audience.

Exercises

Additional Activities related to designing documents and visuals are located at **college.cengage .com/pic/ kolinconcise2e**.

1. Find an ineffectively designed document—a form, a set of instructions, a brochure, a section of a manual, a story in a newsletter—and assume that you are a document design consultant. Write a sales letter to the company or agency that prepared and distributed the document, offering to redesign it and any other documents they have. Stress your qualifications and include a sample of your work. You will have to be convincing and diplomatic—precisely and professionally persuading your readers that they need your services to improve their corporate image, customer relations, and sales or services.

2. Redesign the handwashing document on the next page to make it conform to the guidelines specified in this chapter.

3. One government agency supplied statistics on the world production of oranges (including tangerines) in thousands of metric tons for the following countries during the years 2005–2008: Brazil, 2,005, 2,132, 2,760, 2,872; Israel, 909, 1,076, 1,148, 1,221; Italy, 1,669, 1,599, 1,766, 1,604; Japan, 2,424, 2,994, 2,885, 4,070; Mexico, 937, 1,405, 1,114, 1,270; Spain, 2,135, 2,005, 2,179, 2,642; and the United States, 7,658, 7,875, 7,889, 9,245. Prepare a table with that information and then write a paragraph in which you introduce and refer to the table and draw conclusions from it.

4. Write a paragraph introducing and interpreting the following table.

Year	Soft Drink Companies	Bottling Plants	Per Capita Consumption (Gallons)
1940	750	750	10.3
1945	578	611	12.5
1950	457	466	18.6
1955	380	407	17.2
1960	231	292	15.9
1970	171	229	15.4
1975	118	197	16.0
1980	92	154	18.7
1985	54	102	21.1
1990	43	88	23.1
1995	45	82	25.3
2000	37	78	27.6
2005	34	72	30.1

5. According to a municipal study in 2008 the distribution of all companies classified in each enterprise industry in that city was as follows: minerals, 0.4%; selected services, 33.3%; retail trade, 36.7%; wholesale trade, 6.5%; manufacturing, 5.3%; and construction, 17.8%. Make a circle chart to represent the dis-

WHY SHOULD YOU WASH YOUR HANDS?
Bacteria and viruses (germs) that cause illnesses are spread when you don't wash your hands.
If you don't wash your hands, you risk acquiring:
The common cold or flu
Gastrointestinal illnesses Shigella or hepatitis A
Respiratory illnesses
Should you wash your hands?
You need to wash your hands several times every day. Some important times to wash your hands are:
BEFORE
Preparing or eating food.
Treating a cut wound.
Tending to someone who is sick.
Inserting or removing contacts
After
Using the bathroom.
Changing a diaper or helping a child use the bathroom (don't forget the child's hands)
Handling raw meats/poultry/eggs
Touching pets, especially reptiles
handling garbage
Sneezing or blowing your nose, or helping a child blow his/her nose
Touching any body fluids like blood or mucus
Being in contact with a sick person
Playing outside or with children and their toys
WHEN SHOULD YOU WASH YOUR HANDS?
There is a right way to wash your hands.
Follow these steps and you will help protect yourself and your family from illness.
Like any good habit, proper hand washing must be taught.
Take the time to teach it to your children and make sure they practice.

tribution and write a one- or two-paragraph interpretation to accompany (and explain the significance of) your visual.

6. Make an organizational chart for a business or an agency you worked for recently. Include part-time and full-time employees, but indicate their titles or functions with different kinds of shapes or lines. Then write a brief letter to your employer explaining why this kind of organizational chart should be distributed to all employees. Focus on the types of problems that could be avoided if employees had access to such a chart.

7. Prepare a flow chart for one of the following activities:
   a. jumping a "dead" car battery
   b. giving an injection
   c. making a reservation online
   d. painting a set of louvered doors
   e. checking your credit online
   f. putting out an electrical fire
   g. changing your password

    h. preparing a visual using a graphics software package

    i. any job you do

8. Prepare a drawing of one of the following simple tools and include appropriate callouts with your visual.

    a. PDA   **e.** swivel chair

    b. iPod   **f.** DVD player

    c. pliers   **g.** ballpoint pen

    d. stethoscope   **h.** table lamp

9. Prepare appropriate visuals to illustrate the data listed in a and b. In a paragraph immediately after the visual, explain why the type of visual you selected is appropriate for the information.

    a. Life expectancy is increasing in America. This growth can be dramatically measured by comparing the number of teenagers with the number of older adults (over age 65) in America during the last few years and then project- ing those figures. In 1970 there were approximately 28 million teenagers and 20 million older adults. By 1980 the number of teenagers climbed to 30 million and the number of older adults increased to 25 million. In 1990 there were 27 million teenagers and 31 million older adults. In 2000 the number of teenagers had leveled off to 23 million, but the number of older adults soared to more than 36 million.

    b. Researchers estimate that for every adult in America 3,985 cigarettes were purchased in 1985; 4,100 in 1990; 3,875 in 1995; 3,490 in 2000; and 2,910 in 2005.

10. This visual was prepared to accompany a report on problems pilots have encoun- tered with a particular model of jet engine. Redo the visual to make it easier to read and to organize information. Supply a paragraph to accompany your new visual.

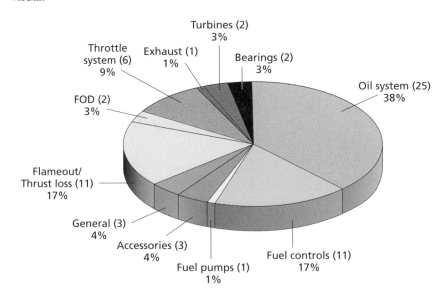

11. Find a misleading visual, from a print or online source, and explain in an e-mail to your instructor why it is deceptive, incomplete, cluttered, inaccurate, and/or biased. Reconstruct the visual and provide a corrected, edited paragraph based upon information in the revised visual. Attach both versions to your e-mail to your instructor.

12. Explain why the following would be inappropriate in communicating with an international audience and how you would revise the document containing such visuals/graphics symbols:
    a. clip art showing a string tied around an index finger
    b. picture of a man with a sombrero in a newsletter for Pronto Check Cashing Company
    c. clip art/drawing of a light bulb and logo "Smart Ideas" for a CPA firm
    d. clip art showing someone crossing the middle finger over the index finger (wishing sign)
    e. drawing of a Cupid figure for a caterer
    f. a sales brochure showing a white and blue flag for a French audience
    g. an ad for a hotel with a picture of the White House.

CHAPTER 7

Writing Instructions and Procedures

To expand your understanding of writing instructions and procedures, take advantage of the Web Links, Additional Activities, and ACE Self-Tests at **college.cengage .com/pic/ kolinconcise2e.**

Clear and accurate instructions are essential to the world of work. Instructions tell—and frequently show—how to do something. They indicate how to perform a procedure (draw blood; change the oil in your car); operate a machine (a pH meter; a digital camera); construct, install, maintain, adjust, or repair a piece of equipment (an incubator; a scanner). Everyone from the consumer to the specialist uses and relies on carefully written and designed instructions.

Instructions and Your Job

As part of your job, you may be asked to write instructions, alone or with a group, for your co-workers as well as for the customers who use your company's services or products. Your employer stands to gain or lose much from the quality and the accuracy of the instructions you prepare.

While the purpose of writing instructions is to explain how to perform a task in a step-by-step manner, the purpose of writing procedures is slightly different. Often the two terms are incorrectly used interchangeably. **Procedures** (or policies) describe not a task to be completed, but a set of established rules of conduct to be followed within an organization, such as a business.

This chapter will first show you how to develop, write, illustrate, and design a variety of instructions, and then move into a discussion of writing procedures (pp. 232–235) about job-related duties.

Why Instructions Are Important

Perhaps no other type of occupational writing demands more from the writer than do instructions because so much is at stake—for both you and your reader. The reader has to understand what you write and perform the procedure as well. You cannot afford to be unclear, inaccurate, or incomplete. Instructions are significant for many reasons, including safety, efficiency, and convenience.

Safety

Carefully written instructions get a job done without damage or injury. Poorly written instructions can be directly responsible for an injury to the person trying to follow them and may result in costly damage claims or even lawsuits. Notice how the product labels in your medicine cabinet inform users when, how, and why to take a medication safely. Without those instructions, consumers would be endangered by taking too much or too little medicine or by not administering it properly.

To make sure your instructions are safe, they must be

- accurate
- consistent
- thorough
- clearly written
- carefully organized

Efficiency

Well-written instructions help a business run smoothly and efficiently. No work would be done if employees did not have clear instructions to follow. Imagine how inefficient it would be for a business if employees had to stop their work each time they did not have or could not understand a set of instructions. Or, equally alarming, what if employees made a number of serious mistakes because of confusing directions, costing a business lost sales and increased expenses? Giving readers helpful tips to make their work easier will also increase their efficiency.

Convenience

Clear, easy-to-follow instructions make a customer's job easier and less frustrating. In the customer's view, instructions reflect a product's quality and convenience. How many times have you heard complaints about a company because the instructions that went with its products were difficult to follow? Poorly written and illustrated instructions will cost you business. Instructions are also a vital part of "service after the sale." Owners' manuals, for example, help buyers to avoid a product breakdown (and the headache and expense of starting over) and to keep it in good working order.

■ The Variety of Instructions: A Brief Overview

Instructions vary in length, complexity, and format. Some instructions are one word long: *stop, lift, rotate, print, erase.* Others are a few sentences long: "Insert blank disk in external disk drive"; "Close tightly after using"; "Store in an upright position."

Instructions can be given in a variety of formats, as Figures 7.1 through 7.3 (pp. 212–214) show. They can be in paragraphs (see Figure 7.5, p. 220), often employ visuals to illustrate each step (Figures 7.1 and 7.3), or use numbered lists (Figures 7.1–7.3). You will have to determine which format is most appropriate for the kinds of instructions you are to write. For writing that affects policies or regulations, as in Figure 7.7 (see pp. 233–234), you will most often use a memo format.

Figure 7.1   Instructions that supply a visual with each step.

Proper Brushing

Proper brushing is essential for cleaning teeth and gums effectively. Use a toothbrush with soft, nylon, round-ended bristles that will not scratch and irritate teeth or damage gums.

1

Place bristles along the gumline at a 45-degree angle. Bristles should contact both the tooth surface and the gumline.

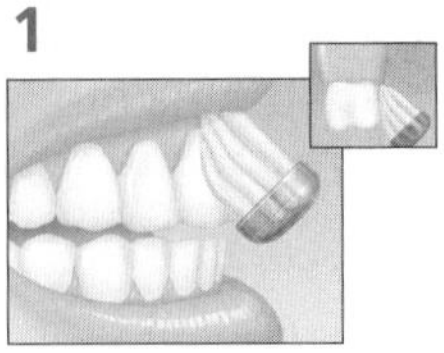

2

Gently brush the outer tooth surfaces of 2–3 teeth using a vibrating back and forth rolling motion. Move brush to the next group of 2–3 teeth and repeat.

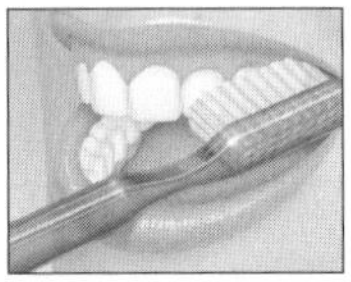

3

Maintain a 45-degree angle with bristles contacting the tooth surface and gumline. Gently brush, using back, forth, and rolling motion along all of the inner tooth surfaces.

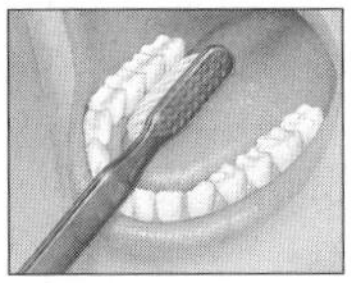

4

Tilt brush vertically behind the front teeth. Make several up and down strokes using the front half of the brush.

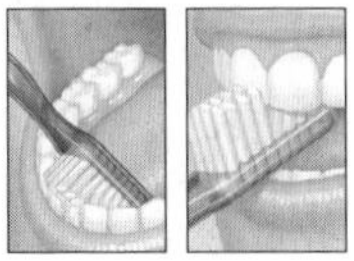

5

Place the brush against the biting surface of the teeth and use a gentle back and forth scrubbing motion. Brush the tongue from back to front to remove odor-producing bacteria.

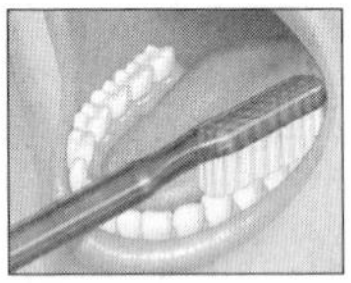

Source: Reprinted by permission of American Dental Hygienists' Association. Illustrations adapted and used courtesy of the John O. Butler Company, makers of *GUM* Healthcare products.

▋Assessing and Meeting Your Audience's Needs

Put yourself in the readers' position. In most instances you will not be available for readers to ask you questions when they do not understand something. Consequently, they will have to rely only on your written instructions.

Instructions given in a numbered list describing a sequence of steps. **Figure 7.2**

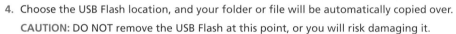

Copying Files to a USB Flash from Your PC or Laptop

1. Insert the USB Flash (see photo) into a USB portal of your PC or laptop.

2. Find the folder or file to be copied to the USB Flash and right click on it.

   NOTE: The folder or file will be highlighted and a menu with "Open" at the top will appear.

3. Within the menu, move your cursor down to the "Send To" option, at which point a list of locations to which you may send the selected folder or file will appear.

4. Choose the USB Flash location, and your folder or file will be automatically copied over.

   CAUTION: DO NOT remove the USB Flash at this point, or you will risk damaging it.

5. Go to "My Computer" from the computer's "Start" menu, and double click on the USB Flash. You will now see if the copying was successful.

6. Eject the USB Flash before removing it from the computer. To do so, go to "My Computer" again, right click on the USB Flash and select the "Eject" option from the menu.

7. Remove the USB Flash from the USB portal.

Do not assume that members of your audience have performed the process before or have operated the equipment as many times as you have. (If they had, there would be no need for your instructions.) No writer of instructions ever disappointed readers by making directions too clear or too easy to follow. Keep in mind that your audience will often include non-native speakers of English, a worldwide audience of potential consumers.

Key Questions to Ask About Your Audience

To determine your audience's needs, ask yourself the following questions:

- How and why will my readers use my instructions? (Engineers have different expectations than do office personnel and customers.)
- What language skills do they possess—is English their first (native) language?
- How much do my readers already know about the product or procedure?
- How much background information will I have to supply?
- What steps will most likely cause readers trouble?
- How often will they refer to my instructions—every day or just as a refresher?
- Where will my audience most likely be following my instructions—in the workplace, outdoors, in a workshop equipped with tools, or alone in their homes?
- What resources, such as special equipment or power sources, will readers need to perform my instructions successfully?

Figure 7.3   Instructions in a numbered list on how to assemble an outdoor grill.

ASSEMBLY INSTRUCTIONS

The instructions shown below are for the basic grill with tubular legs. If you have a pedestal grill, or a grill with accessories, check the separate instruction sheet for details not shown here.

NOTE: Make sure you locate all the parts before discarding any of the packaging material.

TOOLS REQUIRED . . . A standard straight blade screwdriver is the only tool you need to assemble your new Meco grill. If you have a pedestal grill, you will need a 7-16 wrench or a small adjustable wrench.

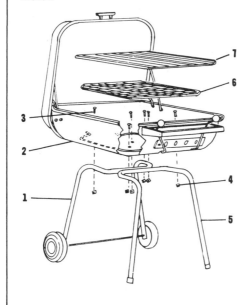

1. Before you start, take time to read through this manual. Inside you will find many helpful hints that will help you get the full potential of enjoyment and service from your new Meco grill.

2. Lay out all the parts.

3. Assemble roller leg (1) to bottom rear of bowl (2) with 1¼″ long bolts (3) and nuts (4).

4. Assemble fixed leg (5) to bottom front of bowl (2) with 1¼″ long bolts (3) and nuts (4).

5. Place fire grate—ash dump (6) in bottom of bowl (2) between adjusting levers.

6. Place cooking grid (7) on top of adjusting levers. Make sure top grid wires run from front to back of grill.

Source: Meco Assembly Instructions and Owners Manual, Metals Engineering Corp., P.O. Box 3005, Greenville, TN 37743. Reprinted by permission.

■ The Process of Writing Instructions

As we saw in Chapter 2, clear and concise writing requires you to follow a process. To make sure your instructions are accurate and easy for your audience to perform, follow these steps.

Plan Your Strategy

Before writing, do some research to understand completely the job, process, or procedure that you are asking someone else to perform. Make sure you know

- the reason for doing something
- the parts or tools required

- the steps to follow to get the job done
- the results of the job
- the potential risks or dangers

If you are not absolutely sure about the process, ask an expert for a demonstration. Do some background reading and talk to or e-mail colleagues who may have written or followed a similar instruction.

Do a Trial Run

Actually perform the job (assembling, repairing, maintaining, dissecting) yourself or with all your writing team present. Go through a number of trial runs. Take notes as you go along and be sure to divide the job into simple, distinct steps for readers to follow. Don't give readers too much to do in any one step. Each step should be **complete, sequential, reliable, straightforward,** and **easy** for your audience to identify and perform.

Write and Test Your Draft

Transform your notes into a draft (or drafts) of the instructions you want readers to follow. Test your draft(s) by asking someone from the intended audience (consumers, technicians) who may never have performed the job to follow your instructions as you have written them. Observe where the individual runs into difficulty—cannot complete or seems to miss a step, gets a result different from yours.

Revise and Edit

Based on your observations and user feedback, revise your draft(s) and edit the final copy of the instructions that you will give to readers. Always consider whether your instructions would be easier to accomplish if you included visuals.

Analyzing the needs and the background of your audience will help you to choose appropriate words and details. A set of instructions accompanying a chemistry set would use different terminology, abbreviations, and level of detail than would a set of instructions a professor gives a class in organic chemistry.

> **General Audience:** Place 8 drops of vinegar in a test tube with a piece of limestone about the size of a pea.
> **Specialized Audience:** Place 8 gtts of CH_3COOH in a test tube and add 1 mg of CO_3.

The instructions for the general audience avoid the technical abbreviations and symbols the specialized audience requires. If your readers are puzzled by your directions, you defeat your reasons for writing them.

▌Using the Right Style

To write instructions that readers can understand and turn into effective action, observe the following guidelines.

1. Use verbs in the present tense and imperative mood. Imperatives are commands that have deleted the pronoun *you*. Note how the instructions in Figures 7.1 through 7.4 contain imperatives—"Move the arrow" instead of "You move the arrow." In instructions, deleting the *you* is not discourteous, as it would be in a business letter or report. The command tells readers, "These steps work, so do them exactly as stated." Choose action verbs such as those listed in Table 7.1.

2. Write clear, short sentences in the active voice. Keep sentences short and uncomplicated. Sentences under twenty words (preferably under fifteen) are easy to read. Note that the sentences in Figures 7.1 through 7.4 are, for the most part, under fifteen words.

3. Use precise terms for measurements, distances, and times. Indefinite, vague directions leave users wondering whether they are doing the right thing. The following vague direction is better expressed through precise revision.

> **Vague:** Turn the distributor cap a little. (*How much is a little?*)
> **Precise:** Turn the distributor cap one quarter of a rotation.

4. Use connective words as signposts. Connective words specify the exact order in which something is to be done (especially when your instructions are written in paragraphs). Words, such as *first, then, before*, help readers stay on course, reinforcing the sequence of the procedures.

Figure 7.4   Exploded drawing showing how to assemble an industrial extension cord.

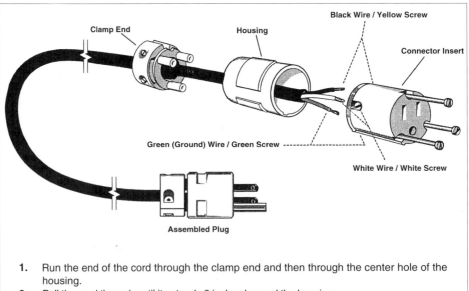

1. Run the end of the cord through the clamp end and then through the center hole of the housing.
2. Pull the cord through until it extends 2 inches beyond the housing.
3. Strip about 1¼ inches of outer insulation from the end of the cord.
4. Twist the exposed ends to prevent stray strands.

Source: Drawing courtesy of Sally Eddy and Georgia-Pacific Company.

TABLE 7.1   Some Helpful Imperative Verbs Used in Instructions

add	determine	insert	pick up	run	tie
adjust	dig	inspect	point	save	tighten
apply	download	install	pour	scan	tilt
blow	drag	lift	press	scroll	trace
call up	drain	load	prevent	select	transect
change	drill	loosen	print	set	transfer
check	drop	lower	pull	send	trim
choose	ease	lubricate	push	shift	turn
clean	eliminate	measure	raise	shut off	twist
click	enter	mix	reboot	slide	unplug
clip	flip	mount	release	slip	use
close	flush	move	remove	spread	ventilate
connect	forward	notify	reply	squeeze	verify
copy	group	oil	review	start	wash
create	hold	open	roll	switch	wind
cut	include	paste	rotate	tear	wipe
delete	increase	peel	rub	thread	wire

5. Number each step when you present your instructions in a list. You also can use bullets. Leave plenty of white space between steps to separate them distinctly for the reader.

Using Visuals Effectively

Readers welcome visuals in almost any set of instructions. Visuals are graphic and direct, helping readers to understand and to see what they must do. A visual can

- simplify a process
- identify the location and size of a part
- show the relationships among components
- reinforce or even save words
- illustrate the "right" way and the "wrong" way
- increase readers' confidence
- help get a job done more quickly

The number and kinds of visuals you include will, of course, depend on the procedure or equipment you are explaining and your audience's background and needs. Some instructions may require only one or two visuals. The instructions in Figure 7.2 show users what they can expect to see on a screen as they download a visual. The shot of the screen clarifies the procedure and assists the reader. In Figure 7.1 each step is accompanied by a visual demonstrating a proper technique of brushing teeth.

Another frequently used visual in instructions is an exploded drawing, like the one in Figure 7.3, which helps consumers see how the various parts of the grill fit together, or the one in Figure 7.4 (p. 216), which labels and shows the relationship of the parts of an industrial extension cord.

Guidelines for Using Visuals in Instructions

Follow these guidelines to use visuals effectively in your instructions:

1. Place the visual next to the step it illustrates, not on another page or buried at the bottom of the page.
2. Assign each visual a number (Figure 1, Figure 2) and refer to visuals by figure number in your instructions.
3. Make sure the visual looks like the object the user is trying to assemble, maintain, run, or repair. Don't use a photo that shows a different model.
4. Always inform readers if a part is missing or reduced in your visual.
5. Where necessary, label or number parts of the visual, as in Figure 7.4.
6. Set visuals off with white space so it is easy to find and examine them.

▌ The Five Parts of Instructions

Except for very short instructions, such as those illustrated in Figures 7.1 through 7.4, or for instructions on policies and regulations (see Figure 7.7, pp. 233–234), a set of instructions generally contains five main parts: (1) an introduction; (2) a list of equipment and materials; (3) the actual steps to perform the process; (4) warnings, cautions, and notes; and (5) a conclusion (when necessary).

Introduction

The function of an introduction is to provide readers with enough *necessary* background information to understand why and how your instructions work. An introduction must make readers feel confident and well prepared before they turn to the actual steps.

What to Include in an Introduction

You can do one or all of the following in your introduction. Not every introduction to a set of instructions will contain facts in all four categories of information listed here. Some instructions will require less detail. You will have to judge how much background information to give readers for the specific instructions you write.

1. State why the instructions are useful for a specific audience. Many instructions begin with introductions that stress safety, educational, or occupational benefits. Here is an introduction from a safety procedure describing protective lockout of equipment.

> The purpose of this procedure is to provide plant electrical technicians with a uniform method of locking out machinery or equipment. This will prevent the possibility of setting moving parts in motion, energizing electrical lines; or opening valves while repair, setup, or cleaning work is in progress.

2. Indicate how a particular machine, procedure, or process works. An introduction can briefly discuss the "theory of operation" to help readers understand why something works the way your instructions say it should. Such a discussion sometimes describes a scientific law or principle. An introduction to instructions on how to run an autoclave begins by explaining the function of the machine: "These instructions will teach you how to operate an autoclave, which is used to sterilize surgical instruments through live additive-free stream."

3. Point out any safety measures or precautions a reader may need to be aware of. By alerting readers early in your instructions, you help them to perform the procedure much more safely and efficiently. Note how Cliff Burgess opens his instructions in Figure 7.5 (pp. 220–221) with a warning about safety.

4. Stress any advantages or benefits the reader will gain by performing the instructions. Make the reader feel good about buying or using your product by explaining how it will make a job easier to perform, save the reader time and money, or allow the reader to accomplish a job with fewer mistakes or false starts. Note how the following introduction to a set of instructions on using an auto dial/auto answer modem encourages the reader to want to learn how to operate this system.

Welcome to high-speed telecommunications and congratulations on choosing the Signalman EXPRESSi. You've made an excellent choice. The EXPRESSi is the ideal link between your computer and the ever-expanding world of information utilities, databases, electronic mail, bulletin boards, computer time sharing, and more.

The EXPRESSi can be used in the IBM Personal Computer. And because the EXPRESSi fits inside your PC, it saves valuable desk space and eliminates expensive, bulky cables.[1]

List of Equipment and Materials

Clearly, some instructions, as in Figure 7.1, do not need to inform readers of all equipment or materials they will need. But when you do, make your list complete and clear. Do not wait until the readers are actually performing one of the steps to tell them that a certain type of drill or a specific kind of chemical is required. They may have to stop what they are doing to find the equipment or material; moreover, the procedure may fail or present hazards if users do not have the right equipment at the right time. For example, if a Phillips screwdriver is essential to complete one step, specify that type of screwdriver under the heading "Equipment and Materials"; do not list just "screwdriver."

---

[1]Courtesy of Anchor Automation, Inc., Chatsworth, CA.

Figure 7.5   An instructional memo listing safety precautions for a general audience.

BURTON WORLDWIDE SYSTEMS
www.burton.com

TO: All Burton Employees
FROM: Cliff Burgess, Environmental Safety C.B.
RE: Video Display Terminal (VDT) Safety Precautions
DATE: September 12, 2008

Introduction emphasizes reasons for instructions

You may experience some possible health risks in using your computer video display terminal (VDT). These risks include sleep disorders, behavioral changes, danger to the reproductive system, and cancer.

Nontechnical explanation offers an analogy

The source of any risks comes from the electromagnetic fields (EMFs) that surround anything that carries an electric current—for example, copiers, circuit breakers, and especially VDTs. Your computer monitor is a major source of EMFs. Magnetic fields can go through walls as easily as light goes through glass.

Although EMFs may affect your health, you can considerably reduce your exposure to these fields by following these three simple steps:

Steps stand out through bullets, bold-face, and spacing

- **Stay at least three feet (an arm's length) away from the front of your VDT.** (The magnetic field is significantly reduced with this amount of distance.)

- **Stay at least four feet away from the sides and back of someone else's VDT.** (The fields are weaker in the front of the VDT but much stronger everywhere else.) Refer to the following drawing, which you may want to post in your office.

Visual clarifies instructions

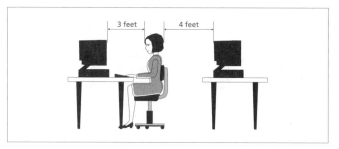

1215 Madison • St. Louis, Missouri 63174
314-555-4300 • FAX 314-555-4311

Continued

(Continued)   Figure 7.5

> 2.
>
> - **Switch your VDT/computer off when you are not using it.** (If the computer has to remain on, be sure to switch off the monitor; screen savers do not affect the exposure to **EMFs**.)
>
> Our environmental safety team will continue to monitor and investigate any problems. Observing the guidelines above, however, will help you to take all the necessary precautions in order to minimize your exposure to EMFs.
>
> If you have any questions about these procedures or your exposure to EMFs, please e-mail me at cliffb@burton.com.

Conclusion reassures readers they are acting safely by following instructions

Source: Reprinted by permission.

Steps for Your Instructions

The heart of your instructions will consist of clearly distinguished steps that readers must follow to achieve the desired results. Figure 7.6 (pp. 225–231) contains a model set of steps on how to set up a printer. Note how each step is precisely keyed to the visual, further helping readers perform the procedure. Refer to this figure as you study this section. To make sure that what you write will help your readers understand your steps, observe the following rules.

1. Put the steps in their correct order and number them. If a step is out of order or is missing, the entire set of instructions can be wrong or, worse yet, dangerous. Double-check every step and number each step to indicate its correct place in the sequence of events you are describing.

2. Put only the right amount of information in each step. Giving readers too little information can be as risky as giving them too much. Keep in mind that each step asks readers to perform a single task in the entire process. However, see the next rule for the exception.

3. Group closely related activities into one step. Sometimes closely related actions belong in one step to help the reader coordinate activities and to emphasize their being done at the same time, in the same place, or with the same equipment.

Instructions on how to use a fax machine are clearer when distinct steps are stated separately. The first set of instructions below incorrectly tells users how to transmit a fax by combining steps that must be performed separately. Step 2 asks users to pick up the phone and then dial the number—two separate actions. Step 3 asks users to press the button and hang up the handset, again two actions that cannot be performed simultaneously.

Incorrect: 1. Load the paper into the outgoing document slot, adjusting the paper guides to the appropriate width.
2. Pick up the telephone handset and listen for a dial tone. When you hear the dial tone, dial the number of the receiving fax machine.
3. When the receiving fax machine answers the ring, press the start button. After the transmission is completed, return the handset to its cradle.

Correct: 1. Load the paper into the outgoing document slot, adjusting the paper guides to the appropriate width.
2. Pick up the telephone handset and listen for a dial tone.
3. When you hear a dial tone, dial the number of the receiving fax machine.
4. When the receiving fax machine answers the ring, press the start button.
5. After the transmission is completed, return the handset to its cradle.

Don't divide an action into two steps if it has to be done in one. For example, instructions showing how to light a furnace would not list as two steps actions that must be performed simultaneously to avoid a possible explosion.

Incorrect: 1. Depress the lighting valve.
2. Hold a match to the pilot light.

Correct: 1. Depress the lighting valve while holding a match to the pilot light.

Similarly, do not separate two steps of a computer command that must be performed simultaneously.

Incorrect: 1. Press the CONTROL key.
2. Press the ALT key.

Correct: 1. While holding down the CONTROL key, press the ALT key.

4. Give the reader hints on how best to accomplish the procedure. Obviously, you cannot do that for every step, but if there is a chance that the reader might run into difficulties you should provide assistance. Particular techniques on how to operate or service equipment also help readers: "If there is blood on the transducer diaphragm, dip the transducer in a blood solvent, such as hydrogen peroxide, Hemosol, etc." If readers have a choice of materials or procedures in a given step, you might want to list those that would give the best performance: "Several thin coats will give a better finish than one heavy coat."

5. State whether one step directly influences (or jeopardizes) the outcome of another. Because all steps in a set of instructions are interrelated, you could not (and should not have to) tell readers how every step affects another. But stating specific relationships is particularly helpful when dangerous or highly intricate operations are involved. You will save the reader time, and you will stress the need for care. Forewarned is forearmed. Here is an example.

Step 2: Tighten fan belt. Failure to tighten the fan belt will cause it to loosen and come off when the lever is turned in step 5.

Do not wait until step 5 to tell readers that you hope they did a good job in tightening the fan belt in step 2. Information that comes after the fact is not helpful.

6. Where necessary, insert graphics to assist readers in carrying out the step. For example, see the drawings of the printer in Figure 7.6 (pp. 225–231).

Warnings, Cautions, and Notes

At appropriate places in the steps of your instructions you may have to stop the reader to issue a warning, a caution, or a note. Here are some examples. These are found in Figure 7.6, especially in step 4.

Warnings

A warning tells readers that a step, if not prepared for or performed properly, can endanger their safety, as here. Note that translations ensure readers' safety.

WARNING: UNPLUG MACHINE BEFORE REMOVING PLATEN GLASS.

ADVERTENCIA: DESENCHUFE LA MAQUINA ANTES DE QUITAR EL VIDRIO.

Spanish translation

Cautions

A caution tells a reader how to avoid a mistake that could damage equipment or to take certain precautions—"Wear protective goggles"; "Do not force the plug."

Caution: Formatting erases all data on the disk

小心：格式化会删除磁盘的所有资料

Chinese translation

Notes

A note adds a clarification, provides a helpful hint on how to do the step most efficiently, or lists different options.

At 20 degrees F, a battery uses about 68 percent of its power.

Eksi 7 derecede, bir pil enerjisinin yaklasik yuzde 68 'ini kullanir.

Turkish translation

Guidelines on Using Warnings, Cautions, and Notes

1. **Do not regard warnings and cautions as optional.** They are vital for legal and safety reasons to protect lives and property. In fact, you and your company can be sued if you fail to notify the users of your product or service of potentially dangerous conditions that will or could result in injury or death.

2. **Put warnings and cautions in the right place.** Place them immediately before the step to which they pertain. If you insert a warning or caution statement too early, readers may forget it by the time they come to the step to which it applies. And if you put the notification too late, you almost certainly expose the reader to danger and the equipment to breakdown.

3. **Put warnings and cautions in a distinctive format.** Warnings and cautions should be graphically set apart from the rest of the instructions, as the triangle on page 223 illustrates. There should be no chance that readers will overlook them. Put such statements in capital letters, boldface type, boxes, different colors (red is especially effective for warnings if your readers are native speakers of English, but see page 204; yellow is often used for caution).

   Be careful, though, about using colors for non-native speakers of English. Use one or all of those devices. Make sure your audience understands an icon or a symbol, like a skull and crossbones, an exclamation point inside a triangle, or a traffic stoplight, often used to signal a warning, a hazard, or some other unsafe condition.

4. **Include enough explanation to help readers know what to watch out for and what precautions to take.** Do not just insert the word WARNING or CAUTION. Explain what the dangerous condition is and how to avoid it. Look at the examples of warnings and cautions in Figure 7.6.

5. **Do not include a warning or a caution just to emphasize a point.** Putting too many in your instructions will decrease the dramatic impact they should have on readers. Use them sparingly—only when absolutely necessary—so readers will not be tempted to ignore them.

6. **Use notes only when the procedure calls for them and they will help readers**, as in Figure 7.3.

Conclusion

Not every set of instructions requires a conclusion. For short instructions containing a few simple steps, such as those in Figures 7.1 through 7.3, no conclusion is necessary. These instructions usefully end with the last step the reader must perform. For longer, more involved jobs, a conclusion can help readers finish the job with confidence and accuracy.

When they are necessary, conclusions can provide a succinct wrap-up of what the reader has done or end with a single sentence of congratulations, or reassure readers as the conclusion in Cliff Burgess's memo (Figure 7.5) (pp. 220–221) does. A conclusion might also tell readers what to expect once a job is finished, describe the results of a test, or explain how a piece of equipment is supposed to operate.

▌Model of Full Set of Instructions

Study Figure 7.6, which is a set of instructions on setting up an Epson printer that includes the parts discussed in this chapter: an introduction; a list of materials; numbered steps; and warnings, cautions, and notes. Pay special attention to how the writer coordinates words with visuals to assist readers.

Complete set of instructions with steps, visuals, cautions, notes, warnings.   **Figure 7.6**

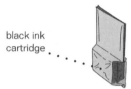

Quick Setup

Here's how to set up your new EPSON® printer . . .

1 Unpack the Printer

Remove any packing material from the printer, as described on the Notice Sheet in the box. Save all the packaging so you can use it if you need to transport the printer later. You'll find these items inside:

black ink
cartridge

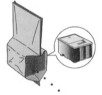

color ink
cartridge

paper
support

Appropriately brief introduction for this straightforward procedure

Each step begins with imperative verb

List (with visuals) of materials

Exploded drawing assists readers in identifying/ assembling parts

Continued

Figure 7.6   (Continued)

Bulleted list highlights importance of correct printer placement

Groups clearly distinguished steps in the procedure

Uses appropriate icon to alert reader to possible injury or damage

Place the printer flat on a stable desk near a grounded outlet. Leave plenty of room in back for the cables and enough room in front for opening the output tray.

Do NOT put the printer:

- In an area with high temperature or humidity
- In direct sunlight or dusty conditions
- Near sources of heat or electromagnetic interference, such as loudspeakers or cordless telephone base units.

Also, be sure to follow the Safety Instructions in the Introduction of your *User's Guide*.

2 Attach the Paper Support

Insert the paper support in the top slot on the back of the printer.

3 Plug In the Printer

First make sure the power is off. Check the ⏻ power button; it's off when its surface is raised above the printer surface.

Caution:
Do not plug the printer into an outlet controlled by a wall switch or timer, or on the same circuit as a large appliance. This may disrupt the power, which can erase memory and damage the power supply.

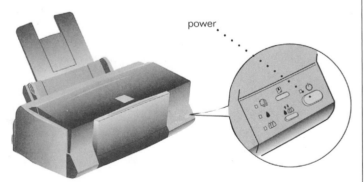

Plug the power cord into a properly grounded outlet.

Continued

(Continued)   **Figure 7.6**

Install the Ink Cartridges

1. Lower the output tray and raise the printer cover.

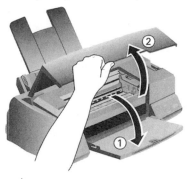

Directional arrows visualize correct movement of parts

2. Press the ⏻ power button to turn on the printer. The ⏻ power light flashes, the ♦ and 🌢 ink out lights come on, and the ink cartridge holders move to the installation position.

3. Pull up the ink cartridge clamps.

Visual clues reinforce right way to perform procedure

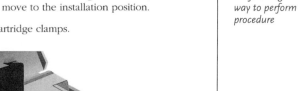

Caution:

You must remove the tape seal from the top of the cartridge or you will permanently damage it. Don't remove the tape seal from the bottom or ink will leak.

4. Open the ink cartridge packages. Remove the disposable yellow portion of the tape seal on top.

Caution and warning notices inserted in right places with distinctive icons

Warning:

If ink gets on your hands, wash them thoroughly with soap and water. If ink gets in your eyes, flush them immediately with water.

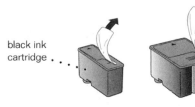

black ink cartridge . . . . .

color ink cartridge

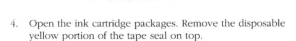

Continued

Figure 7.6   (Continued)

Callouts make parts easy to find

Generous white space makes steps easy to distinguish and follow

Appropriate icon alerting reader of a possible problem

5. Lower the ink cartridges into their holders with the labels face up and the arrows pointing toward the back of the printer.

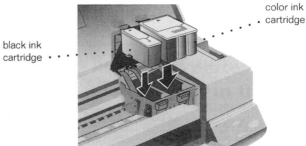

color ink cartridge

black ink cartridge

6. Push down the clamps until they lock in place.

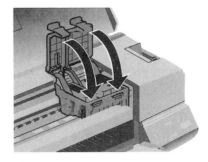

Caution:
Never turn off the printer when the ☉ power light is flashing.

7. Press the ● cleaning button to return the print heads to their home position and charge the ink delivery system. Charging can take up to five minutes, with the ☉ power light flashing until it's finished.

8. Close the printer cover.

Continued

(Continued)   **Figure 7.6**

Load the Paper

1. Slide the left edge guide all the way left and pull out the output tray extension.

Includes two visuals, one of them an exploded drawing, helpfully inserted between the steps to which they apply

2. Fan a stack of plain paper and then even the edges.

3. Load the stack with the printable surface face up. Push the paper against the right edge guide.

Offers clear-cut directions on how to load paper

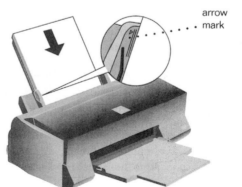

arrow
· · · · · mark

Note:
Don't load paper above the arrow mark inside the left edge guide.

Helpful information to ensure best use of equipment

4. Slide the left edge guide back against the stack of paper.

Continued

Figure 7.6   (Continued)

6 Check the Printer

1. Turn off the printer.

2. While holding down the ⏏ load/eject button, turn on the printer. Then release the buttons.

3. A page prints out showing the ROM version and a nozzle check pattern. When it's finished, turn off the printer. If you have any problems with the test, see Chapter 6 in your *User's Guide* for more information.

Provides a useful source for further information on performing step correctly

7 Connect the Printer to Your Computer

You can connect your EPSON Stylus™ COLOR 600 to either an IBM® compatible PC or an Apple® Macintosh® You'll need a shielded, twisted-pair parallel cable to connect to a PC or an Apple System Peripheral-8 cable to connect to a Macintosh. For a complete list of system requirements, see the Introduction of your *User's Guide*.

Inserts a brief introduction that alerts users to follow one of two procedures, depending on the type of computer they have

Connecting to a PC

1. Turn off the printer and your computer.

Chooses informative icon for a note

Note:
The printer is assigned to parallel port LPT1; if you want to use a different port, see your Windows documentation for instructions.

2. Connect the cable to the printer's parallel interface; then squeeze the wire clips together until they lock in place. (If your cable has a ground wire, connect it now.)

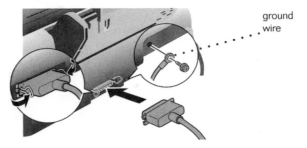

ground wire

Enlarged view of highlights helps readers see step

3. Connect the other end of the cable to your computer's parallel port and secure it as necessary.

Continued

Connecting to a Macintosh

1. Turn off the printer and your Macintosh.

2. Connect one end of the cable to the serial connector on the back of the printer.

Visual identifies precise location for connecting parts

Note:
If you're using a
PowerBook™, connect your
printer to the modem port.

3. Connect the other end of the cable to either the modem port ✆ or the printer port 🖶 on your Macintosh.

Install the Printer Software

Now you need to install the printer software so you can control printing from your computer.

Functions as both a step—installing software—and a wrap-up

Installing on a PC

You can install the printer software for Windows 95 or Windows 3.1 from the EPSON printer software CD-ROM. If you don't have a CD-ROM drive, you can install the software using the EPSON printer software diskettes.

Installing from the CD-ROM

In addition to the printer driver and utilities, the CD-ROM contains EPSON Answers, a comprehensive online guide that includes:

EPSON ANSWERS
If you have a CD-ROM
drive, you can run EPSON
Answers, the on-screen
guide to your new printer.
It puts you on the right
track quickly and easily.

▶ **How To** for step-by-step printer operating instructions
▶ **Color Guide** with practical color printing information
▶ **Problem Solver** to help you fix printer problems
▶ **Test Print** so you can check your print quality

Follow the instructions inside the CD-ROM case to install the software. To run EPSON Answers, click on its icon in the EPSON program group or folder.

Provides information for further troubleshooting and contact with manufacturer

Source: Reprinted by permission of Epson America.

Writing Procedures for Policies and Regulations

Up to this point, we have concentrated primarily on instructions dealing with how to put things together; how to install, repair, or use equipment; and how to alert readers to mechanical or even personal danger(s). But there is another similar type of writing that deals with guidelines for the world of work: procedures. These concern policies and regulations found in employee handbooks and other internal corporate communications, such as on websites, in memos, or in e-mail messages. (Note, however, that some companies do not disseminate policy matters via e-mail because it is perceived as less formal than hard copy and is not deletable.) Figure 7.7 shows an example of one company's policy on flextime.

Some Examples of Procedures to Write

Procedures deal with a wide range of activities within an organization, including the following:

- accessing a company file or database
- preparing for an audit
- applying for family leave
- dressing professionally at work
- routing information
- reserving a company vehicle or facility
- taking advantage of telecommuting options
- filing a work-related grievance
- requesting travel expense reimbursement
- fulfilling promotion requirements

Policy procedures have a major impact on a company and its workers; they affect schedules, payrolls, acceptable and unacceptable behaviors at work, and a range of protocols governing the way an organization does business internally and externally.

Meeting the Needs of Your Marketplace

As with instructions, you will have to plan carefully for writing procedures. A mistake in business procedures can be as wide ranging and as costly as an error in a set of assembly instructions—perhaps more so since poorly written procedures can land a company and/or its employees in significant legal trouble.

To avoid such difficulties, spell out precisely what is expected of employees—how, when, where, and why they are to perform or adhere to a certain policy. Use the same strategies as for instructions discussed earlier in this chapter. Leave no chance for misunderstanding or ambiguity; be unqualifyingly straightforward and clear-cut. Determine what information employees need to comply with all the regulations and make sure you supply it.

Instructions on a new scheduling policy.   **Figure 7.7**

WebTech, Inc.   www.webtech.com
4300 Ames Boulevard, Gunderson, CO 81230-0999   (303) 555-9721
                                                     FAX (303) 555-9876

TO: All Employees T.B.   DATE: March 21, 2008
FROM: Tequina Bowers   SUBJECT: Opting for a Flextime
          Human Resources Director   Schedule

Effective sixty days from now, on May 10, 2008, employees will have the
opportunity to go to a flextime schedule or to remain on their current 8-hour
fixed schedule. This memo explains the flextime option and establishes the
guidelines and rules you must follow if you choose this new schedule.

Notifies readers of new policy and states purpose of the memo

Flextime Defined

Flextime is based on a certain number of **core hours** and **flexible hours**. Our
company will be open 12 hours, from 6:00 a.m. to 6:00 p.m. weekdays, to
accommodate both fixed and flextime arrangements. During this 12-hour period,
all employees on flextime will be expected to work **eight consecutive hours,**
including one hour for lunch. All employees must work a **common core time
from 10:00 a.m. to 4:00 p.m.** but flextime employees will be free to choose
their own starting and quitting times. For instance, they might elect to arrive at
8:00 a.m. and leave at 4:00 p.m., or they may want to start at 9:30 a.m. and leave
at 5:30 p.m.

Spells out precisely how company defines flextime; boldfaces most important information

Flextime Guidelines and Rules

Employees need to understand their individual responsibilities and adjust their
schedules accordingly. All employees must adhere strictly to the following
regulations and realize that flextime schedule privileges will be revoked for
violations.

Stresses employee responsibilities and consequences of violating rules

Continued

Figure 7.7   (Continued)

2.

Carefully outlines what is and is not acceptable according to new policy

What Flextime Employees Must Do

(1) Be present during core time, arriving and leaving the plant during their flexible work hours.
(2) Observe a minimum unpaid one hour lunch break each working day.
(3) Cooperate with their supervisors to make sure coverage is provided for their departments from 6:00 a.m. to 6:00 p.m.
(4) Notify supervisors of absences.
(5) Attend monthly corporate meetings even though such meetings may be outside their chosen flextime schedules.
(6) Adhere to dress code during any time they are at work, regardless of their flextimes.
(7) Agree to work on a flextime schedule for a 6-month period.

Numbered points make policy easier to understand, follow, and refer to in the future

What Flextime Employees Can't Do

(1) Be tardy during core time.
(2) Switch, bank, borrow, or trade flextime hours with other employees without the approval of an immediate supervisor.
(3) File for overtime without a supervisor's approval.
(4) Self-schedule a vacation by expanding flextime hours.
(5) Switch back and forth between fixed time and flextime.

Stipulates what new policy will not allow in separate section of policy memo

How Do You Sign Up for Flextime?

If you wish to begin a flextime schedule, first you need to obtain and complete a transfer of hours form from your supervisor. Next, you need to bring that signed form personally to Human Resources (Admin. 201) to participate officially in this program.

Explains steps to begin flextime

I will be happy to talk to you about this new option and to answer any of your questions. Please call me at ext. 5121, e-mail me at **tbowers@webtech.com**, or come by my office in Admin. 201.

Encourages feedback and questions

Thank you.

Many times procedure statements involve a change in the work environment. Help readers by including, whenever necessary, definitions, headings, some prefatory explanations, and an offer to help employees with any questions they may have. Always present a copy of the procedures you draft to management, and to your company's human resource department, to approve or to revise before sending them to employees.

Figure 7.7 shows a memo from a human resources director notifying employees how they can take advantage of a new flextime arrangement at work, including what they can and cannot do within the framework of flextime. Note how the writer divides her procedures into an introduction explaining when flextime will go into effect and what choices employees have about it, a section that usefully defines and delineates the concept of flextime, and, finally, a section containing specific guidelines. These guidelines, while not sequential, function as a series of steps that employees have to follow. Ultimately, these steps will affect the entire organization.

The various regulations about what employees cannot do in flextime might be seen as the equivalents of warning and caution statements discussed in this chapter's section "Warnings, Cautions, and Notes" (pp. 223–224). Observe, too, how the writer does not veer off to discuss benefits to the employer or to examine where and how often flextime has been used elsewhere. Finally, this example of procedural writing protects the employer legally by establishing the policies and rules by which an employee's scheduled work time is clearly defined, delineated, and evaluated.

Some Final Advice

Perhaps the most important piece of advice to leave you with is this: Do not take *anything* for granted when you have to write a set of instructions. It is wrong and on occasion dangerous to assume that your readers have performed the procedure before, that they will automatically supply missing or "obvious" information, or that they will easily anticipate your next step. No one ever complained that a set of instructions was too clear or too easy to follow.

✓ Revision Checklist

- [ ] Analyzed my intended audience's background, especially why and how they will use my instructions.
- [ ] Tested my instructions to make sure they include all necessary steps in their proper sequence.
- [ ] Made sure all measurements, distances, times, and relationships are precise and correct.
- [ ] Avoided technical terms if my audience is not a group of specialists in my field.
- [ ] Used the imperative mood throughout my instructions.
- [ ] Wrote clear, short, active voice sentences.
- [ ] Chose effective visuals, labeled them, and placed them next to the step(s) to which they apply.
- [ ] Made my introduction proportionate to the length and complexity of my instructions and suitable for my readers' needs.

Continued

Continued

- ☐ Included necessary background, safety, and operational information in the introduction.
- ☐ Provided a complete list of tools and materials my audience needs to carry out the instructions.
- ☐ Put instructions in easy-to-follow steps and in the right order.
- ☐ Used numbers or bullets to label the steps and inserted connective words to indicate order.
- ☐ Inserted warnings, cautions, and notes where necessary and made them easy to see and to read.
- ☐ Supplied a conclusion that summarizes what readers should have done or reassures them that they have completed the job satisfactorily.
- ☐ Spelled out clearly responsibilities, benefits, restrictions, and consequences of any procedures for readers.
- ☐ Defined any terms readers may be unfamiliar with in procedures for policies and regulations.
- ☐ Gave a copy of procedures to administrators for their approval before distributing to employees.

Exercises

Additional Activities related to writing instructions and procedures are located at **college.cengage.com/pic/kolinconcise2e**.

1. Find a set of instructions that does not contain any visuals, but that you think should have some graphic material to make it clearer. Design those visuals yourself and indicate where they should appear in the instructions.

2. From a technical manual in your field or in an owner's manual, locate a set of instructions that you think is poorly written and illustrated. In a memo to your instructor, explain why the instructions are unclear, confusing, or badly formatted. Then revise the instructions to make them easier for the reader to carry out. Submit the original instructions with your revision.

3. Write a set of instructions in numbered steps (or in paragraph format) on one of the following relatively simple activities.
   a. tying a shoe
   b. uploading a song from a CD onto your iPod
   c. unlocking a door with a key
   d. sending a text message from a cell phone
   e. planting a tree or a shrub
   f. changing a computer password
   g. removing a stain from clothing
   h. pumping gas into a car
   i. creating a blog

      j. checking a book out of the library
      k. paying with your debit card
      l. shifting gears in a car
     m. checking your e-mail from an Internet cafe

4. Write an appropriate introduction and conclusion for the set of instructions you wrote for Exercise 3.

5. Write a set of full instructions on one of the following more complex topics. Identify your audience. Include an appropriate introduction; a list of equipment and materials; numbered steps with necessary warnings, cautions, and notes; and an effective conclusion. Also include whatever visuals you think will help your audience.
      a. scanning a document
      b. changing a flat tire
      c. testing chlorine in a swimming pool
      d. shaving a patient for surgery
      e. changing the oil and oil filter in a car
      f. surveying a parcel of land
      g. pruning hedges
      h. jumping a dead car battery
      i. using the Heimlich maneuver to help a choking individual
      j. filleting a fish
      k. creating a logo for a letterhead
      l. taking someone's blood pressure
     m. copying a file from a CD onto a hard drive
      n. painting a car
      o. cooking a roast
      p. flossing a patient's teeth after cleaning
      q. downloading an MP3 file from the Web

6. The following set of instructions is confusing, vague, and out of order. Rewrite the instructions to make them clear, easy to follow, and correct. Make sure that each step follows the guidelines outlined in this chapter.

Reupholstering a Piece of Furniture

(1) Although it might be difficult to match the worn material with the new material, you might as well try.
(2) If you cannot, remove the old material.
(3) Take out the padding.
(4) Take out all of the tacks before removing the old covering. You might want to save the old covering.
(5) Measure the new material with the old, if you are able to.
(6) Check the frame, springs, webbing, and padding.
(7) Put the new material over the old.
(8) Check to see if it matches.
(9) You must have the same size as before.

(10) Look at the padding inside. If it is lumpy, smooth it out.

(11) You will need to tack all the sides down. Space your tacks a good distance apart.

(12) When you spot wrinkles, remove the tacks.

(13) Caution: in step 11 directly above, do not drive your tacks all the way through. Leave some room.

(14) Work from the center to the edge in step 11 above.

(15) Put the new material over the old furniture.

P.S. Use strong cords whenever there are tacks. Put the cords under the nails so that they hold.

---

7. Write a set of procedures, similar to Figure 7.7, on one of the following policies or regulations at your school or job site:
   a. offering quality customer service over the phone or via the Internet
   b. filing a claim for a personal injury on the job
   c. designing an employee's personal space—what is and is not allowed?
   d. using the Internet at work for personal use
   e. ensuring confidentiality at work (to practice professional ethics in the workplace)
   f. enrolling in mandatory courses to maintain a license or certificate
   g. playing music in the workplace
   h. going through an orientation procedure (e.g., to begin a new job)
   i. following an acceptable dress code
   j. applying for maternity or sick leave
   k. enrolling in the company's 401K retirement plan
   l. observing the proper care and feeding of pets in a kennel

CHAPTER 8

Writing Effective Short Reports and Proposals

This chapter shows you how to write short reports and proposals, which are among the most important and frequent types of business communications you may be called on to write. Proposals are used to attract new business or recommend changes, while short reports are essentially informational. Both are crucial to day-to-day operations of a company or organization; both can be presented in letter or memo format; and both are designed for a decision-making audience.

To expand your understanding of writing short reports and proposals, take advantage of the Web Links, Additional Activities, and ACE Self-Tests at **college .cengage .com/pic/ kolinconcise2e.**

Why Short Reports Are Important

A short report can be defined as an organized presentation of relevant data on any topic—money, travel, time, technology, personnel, equipment, weather, management—that a company or agency tracks in its day-to-day operations. Short reports are practical and to the point. They show that work is being done, and they also show your boss that you are alert, professional, and reliable. Short reports are written to co-workers, employers, vendors, and clients. When they are intended for individuals within your organization, short reports are most often sent as memos. But for clients usually you will send your reports out as letters.

Businesses cannot function without short written reports. Reports tell whether

- work is being completed
- schedules are being met
- costs have been contained
- sales projections are being met
- unexpected problems have been solved

You may write an occasional report in response to a specific question, or you may be required to write a daily or weekly report about routine activities.

Types of Short Reports

To give you a sense of some of the topics you may be required to write about, here is a list of various types of short reports common in the business world.

appraisal report	inventory report	production report
audit report	investigative report	progress/activity report
budget report	justification report	recommendation report
construction report	laboratory report	research report
design report	manager's report	sales report
evaluation report	medicine/treatment	status report
experiment report	error report	survey report
feasibility report	operations report	test report
incident report	periodic report	travel report

This chapter concentrates on five of the most common types of reports you are likely to encounter in your professional work.

1. periodic reports
2. sales reports
3. progress reports
4. travel reports
5. incident reports

Although there are many kinds of short reports, they all are written for readers who need factual information so that they can get a job accomplished. Never think of the reports you write as a series of casual notes jotted down for *your* convenience.

Anticipate How an Audience Will Use Your Report

Knowing who will read your report and why is crucial to your success as a writer. A co-worker or someone else in your field may be familiar with technical information, such as professional jargon or procedural details. But managers, who will constitute the largest audience for your report, may not always know or be interested in such technical information. Instead, they will want bottom-line details about costs and schedules, for example. Similarly, audiences outside of your company (clients, media personnel, community agencies, etc.) will likely not be interested in technical information. Rather, they want information that helps them understand your company, how it will work with, serve, or participate with them.

All audiences, however, will expect you to do the following in your report writing:

- Begin with the most important information. Do not keep your audience in suspense. Give readers what they want as quickly as possible. Your first sentence should focus on why the report was written and why it is important.
- Write concisely and clearly. Avoid wordiness and repetitions. Steer clear of long paragraphs and hard-to-read sentences. Start with a precise subject line if your report is in a memo e-mail format.
- Supply sufficient and relevant documentation. Don't pad your report with unnecessary details or trivia that will slow down or annoy readers. Focus on how your evidence supports the plan, project, the big picture.

Guidelines for Writing Short Reports

Although there are many kinds of short reports, the following guidelines will help you write any short report successfully.

1. Do the Necessary Research

An effective short report needs the same careful research that goes into other on-the-job writing. Your research may be as simple as instant messaging, e-mailing, or leaving a voice mail for a colleague or inspecting a piece of equipment. Or you may have to test or inspect a product or service or assess the relative merits of a group of competing products or services. Some frequent types of research you can expect to do on the job include:

- checking data in reference manuals or code books
- searching archives and databases for recent discussions of a problem or procedure
- reading background information in professional and trade journals
- reviewing a client's file
- testing equipment
- performing an experiment or procedure
- conferring with colleagues, managers, vendors, or clients
- visiting and describing a site; taking field notes
- attending a conference

Never trust your memory to keep track of all the details that go into making a successful short report. Take notes, either by hand or on your laptop or handheld. A well-kept set of notes will provide you with the basic information you need for your report.

2. Be Objective and Ethical

Your readers will expect you to report the facts objectively and impartially—costs, sales, weather conditions, eyewitness accounts, observations, statistics, test measurements, and descriptions. Your reports should be truthful, accurate, and complete. Here are some guidelines.

- Avoid *guesswork*. If you don't know or have not yet found out, say so and indicate how you'll try to find out.
- Do not substitute *impressions* or *unsupported personal opinions* for careful research.
- Using *biased, skewed,* or *incomplete data* is unethical. Provide a straightforward and honest account; don't exaggerate or minimize.

Review the discussion of ethics in business writing in Chapter 1 (pp. 17–22).

3. Organize Carefully

Organizing a short report effectively means that you include the right amount of information in the most appropriate places for your audience. Many times a simple chronological or sequential organization will be acceptable for your readers. Your headings will show how you organized your report. Readers will expect your report to contain information on the purpose, findings, conclusions, and, in many reports, recommendations, as described in the following sections.

Purpose

Always begin by telling readers why you are writing and by alerting them to what you will discuss and why it is significant. Give your readers a summary of key events and details at the beginning to help them follow the remainder of the report quickly. When you establish the scope (or limits) of your report, you help readers zero in on specific times, places, procedures, or problems.

Findings

This should be the longest part of your report and contain the data you have collected—facts about prices, personnel, equipment, events, locations, incidents, or tests. Gather the data from your research; personal observations; interviews; and/or conversations with co-workers, employers, or clients. Again, remember to choose only those details that are of most importance to your reader.

Conclusion

Your conclusion tells readers what your data means. A conclusion can summarize what has happened; review what actions were taken; or explain the outcome or results of a test, a visit, or a program.

Recommendations

A recommendation informs readers what specific actions you think your company or client should take—market a new product, hire more staff, institute safety measures, select among alternative plans or procedures, and so on. Recommendations must be based on the data you collected and the conclusions you have reached. They need to show how all the pieces fit together.

Note how the periodic report in Figure 8.1 fails to help readers see and understand the organization and importance of the information. The revised version of the report, Figure 8.2 (pp. 244–245), clearly illustrates effective report writing; it is an example of a periodic report. The reader is provided with information at regularly scheduled intervals—daily, weekly, bimonthly (twice a month).

4. Use Reader-Centered Headings, Bullets and Numbering, and Visuals

Help readers locate information in your report quickly by using headings and subheadings that clearly demonstrate how you have organized and subdivided your report. See, for example, Figures 8.2, 8.4, and 8.5. Use bulleted and numbered lists where relevant, as in Figure 8.2, to break up chunks of information into easy-to-grasp pieces. Finally, as we discussed in Chapter 6, include visuals to reinforce and summarize key data for your readers, as the table in Figure 8.2 and the map in Figure 8.5 do.

▉ Periodic Reports

Periodic reports, as their name signifies, provide readers with information at regularly scheduled intervals—daily, weekly, bimonthly (twice a month), monthly, quarterly. They help a company or agency keep track of the quantity and quality of the

An example of a poorly written, poorly organized, and poorly formatted short report.   **Figure 8.1**

To Serve and To Protect

Greenfield Police Department

Emergency 555-1000   **Administration** 555-1001   **Traffic** 555-1002

TO: Capt. Alice Martin
FROM: Sergeants Daniel Huxley, Jennifer Chavez,
                and Ivor Paz
SUBJECT: Crimes
DATE: July 12, 2008

This report will let you know what happened this quarter as opposed to what happened last quarter as far as crimes are concerned in Greenfield. This **report is based on statistics** the department has given us over the quarter.

Here we'll let the facts speak for themselves. From Jan.–Mar. we saw 126 robberies while from Apr.–June we had 106. Home burglaries for this period: 43; last period: 36. 33 cars were stolen in the period before this one; now we have 40. Interestingly enough, **last year at this time we had only 27** thefts. Four of them involved heirlooms.

Homicides were 9 this time versus 8 last quarter; assault and battery charges were 92 this time, 77 last time. Carrying a concealed weapon 11 (10 last quarter). We had 47 arrests (55 last quarter) for charges of possession of **a controlled substance**. Rape charges were 8, **1 less than last quarter.** 319 citations this time for moving violations: speeding 158/98, and failing to observe the signals 165/102 last quarter. DUIs this quarter—only 45, or 23 fewer than last quarter.

Misdemeanors this time: disturbing peace 53; vagrancy/ public drunkenness 8; violating leash laws 32; violating city codes 39, including **dumping trash**. Last quarter the figures were 48, 59, 21, 43.

We believe this report is **complete and up-to-date**. We further hope that this report has given you all the **facts** you will need.

Poor format— some para- graphs indented, some not; boldfacing not helpful

Vague subject line

Introduction doesn't tell reader any- thing about overall picture

Throws facts out without any sense of reader's needs

Irrelevant data

No analysis or guided com- mentary—just undigested numbers

Hard-to-follow comparisons and contrasts

Conclusion provides no summary or recommen- dation

Figure 8.2   A well-prepared quarterly report, revised from Figure 8.1.

To Serve and To Protect

Greenfield Police Department

GPD

Emergency 555-1000   Administration 555-1001   Traffic 555-1002

TO: Captain Alice Martin
FROM: Sergeants Daniel Huxley, Jennifer Chavez, and Ivor Paz
SUBJECT: Crime rate for the second quarter of 2008
DATE: July 12, 2008

From April 1 to June 30, 852 crimes were committed in Greenfield, representing a 5 percent increase over the 815 crimes recorded during the previous quarter.

TYPES OF CRIME
The following report, based on the table below, discusses the specific types of crimes, organized into four categories: **robberies and theft, felonies, traffic,** and **misdemeanors**.

Table 1. Comparison of the 1st and 2nd Quarter Crime Rates in Greenfield

CATEGORY	1st Quarter	2nd Quarter
ROBBERIES and THEFT		
Commercial	63	75
Domestic	36	43
Auto	33	40
FELONIES		
Homicides	16	13
Assault and battery	77	92
Carrying a concealed weapon	10	11
Poss. of a controlled substance	55	47
Rape	9	8
TRAFFIC		
Speeding	165	158
Failure to observe signals	102	98
DUI	78	45
MISDEMEANORS		
Disturbing the peace	48	53
Vagrancy	40	48
Public drunkenness	19	40
Leash law violations	21	32
Dumping trash	35	37
Other	8	12

Robberies and Theft
The greatest increase in crime was in robberies, 20 percent more than last quarter. Downtown merchants reported 75 burglaries, exceeding $985,000. The biggest theft occurred on May 21 at Weisenfarth's Jewelers when three armed robbers stole more than $97,000 in merchandise. (Suspects were

Continued

Page 2

apprehended two days later.) Home burglaries accounted for 43 crimes, though the thefts were not confined to any one residential area. We also had 40 car thefts reported and investigated.

Felonies

Homicides decreased slightly from last quarter—from 16 to 13. Charges for battery, however, increased—15 more than we had last quarter. Arrests for carrying a concealed weapon were nearly identical this quarter to last quarter's total. But the 47 arrests for possession of a controlled substance were appreciably down from the first quarter. Arrests for rape for this quarter also were less than last quarter's. Three of those rapes happened within one week (May 6–12) and have been attributed to the same suspect, now in custody.

Provides essential background and statistical information and comparative analyses

Traffic

Traffic violations for this period were lower than last quarter's figures, yet this quarter's citations for moving violations (335) represent a 5 percent increase over last quarter's (322). Most of the citations were issued for speeding (158) or for failing to observe signals (98). Officers issued 45 citations to motorists for DUIs, an impressive decrease over the 78 DUIs issued last quarter. The new state penalty of withholding a driver's license for six months of anyone convicted of driving while under the influence appears to be an effective deterrent.

Easy-to-read sentences

Misdemeanors

The largest number of arrests in this category were for disturbing the peace—53. Compared to last quarter, this is an increase of 10 percent. There were 88 charges for vagrancy and public drunkenness, an increase from the 59 charges made last quarter. We issued 32 citations for violations of leash laws, which represents a sizable increase over last quarter's 21 citations. Thirty-seven citations were issued for dumping trash at the Mason Reservoir.

Includes only data reader needs

CONCLUSION

Overall, while the crime rate has decreased in traffic (especially DUIs) and possession of controlled substances this quarter, we have seen a marked increase in arrests for robberies and battery.

Summarizes findings of report

RECOMMENDATIONS

To help deter robberies in the downtown area, we recommend the following:

- increasing surveillance units in the area
- offering merchants our workshop on safety and security precautions, as we did during the first quarter

Historically, battery arrests have risen during the second quarter. Our recommendations to counter this trend include:

- continuing to work closely with Neighborhood Watch Groups
- providing more foot and bicycle patrols in the neighborhoods with the highest incidence of battery complaints

Offers specific actions/ changes department should make based on conclusion of report in bulleted lists

services it provides and the amount and types of work done by employees. Information in periodic reports helps managers make schedules, order materials, assign personnel, budget funds, and, generally speaking, determine corporate needs.

Figure 8.2 provides essential police information clearly and concisely. As we saw, it summarizes, organizes, and interprets the data collected over a three-month period from individual officers' activity logs. Because of this report, Captain Alice Martin will be better able to plan future protection for the community and to recommend changes in police services.

Sales Reports

Sales reports provide businesses with a necessary and ongoing record of accounts, on-line and mail purchases, losses, and profits over a specified period of time. They help businesses assess past performance and plan for the future. In doing that, they fulfill two functions: **financial** and **managerial.** As a financial record, sales reports list costs per unit, discounts or special reductions, and subtotals and totals. Like a spreadsheet, sales reports show gains and losses. They may also provide statistics for comparing two quarters' sales.

Sales reports are also a managerial tool because they help businesses make both short- and long-range plans. The restaurant manager's sales report illustrated in Figure 8.3 (p. 247) guides the owners to decide which popular entrées to highlight and which unpopular ones to modify or delete. Note how the recommendations follow logically from the figures Sam Jelinek gives to Gina Smeltzer and Alfonso Zapatta, the owners of The Grill.

Progress Reports

A progress report informs readers about the status of an ongoing project. It lets them know how much and what type of work has been done by a particular date, by whom, how well, and how close the entire job is to being completed. A progress report emphasizes whether you are

- keeping your schedule
- staying within your budget
- using the proper technology/equipment
- making the right assignments
- identifying an unexpected problem or possible glitch
- providing adjustments in schedules, personnel, and so on
- completing the job efficiently, correctly, and according to codes

Almost any kind of ongoing work can be described in a progress report—research for a paper, construction of an apartment complex, preparation of a website, documentation of a patient's rehabilitation. Progress reports are often prepared at key phases, or milestones, in a project.

Audience for a Progress Report

A progress report is intended for people who generally are not working alongside you but who need a record of your activities to coordinate them with other indi-

A sales report to a manager.   **Figure 8.3**

Thegrill

Dayton, OH 43210 • (813) 555-4000 • (813) 555-4100 fax   www.Thegrill.com

TO: Gina Smeltzer   DATE: June 28, 2009
         Alfonso Zapatta, Owners
FROM: Sam Jelinek S J.   SUBJECT: Analysis of entrée sales,
         Manager   June 10–16 and June 17–23

As we agreed at our monthly meeting on June 4, here is my analysis of entrée sales
for two weeks to assist us in our menu planning. Below is a record of entrée sales for
the weeks of June 10–16 and June 17–23 that I have compiled and put in a table for
easier comparisons.

	Portion size	June 10–16 Amount	Percentage	June 17–23 Amount	Percentage	2 weeks combined Amount	Percentage
Cornish Hen	6 oz.	238	17	307	17	545	17
Stuffed Young Turkey	8 oz.	112	8	182	10	294	12
Broiled Salmon Steak	8 oz.	154	11	217	12	371	13
Brook Trout	12 oz.	182	13	252	14	434	9
Prime Rib	10 oz.	168	12	198	11	366	11
Lobster Tails	2–4 oz.	147	10	161	9	308	10
Delmonico Steak	10 oz.	56	4	70	4	126	4
Moroccan Chicken	6 oz.	343	25	413	23	756	24
		1,400	**100**	**1,800**	**100**	**3,200**	**100**

Recommendations

Based on the figures in the table above, I recommend that we do the following:

1. Order at least 100 more pounds of prime rib each two-week period to be eligible for
   further quantity discounts from the Northern Meat Company.
2. Delete the Delmonico steak entrée because of its low acceptance.
3. Introduce a new chicken or fish entrée to take the place of the Delmonico steak; I
   would suggest grilled lemon chicken to accommodate those patrons interested in
   tasty, low-fat, lower-cholesterol entrées.

Please give me your reactions within the next week. It shouldn't take more than a few
days to implement these changes.

viduals' efforts and to learn about problems or changes in plans. For example, since supervisors may not be in the field or branch office or at a construction site, they will rely on your progress report for crucial information. Customers, such as a contractor's clients, expect reports on how carefully their money is being spent. That way they can adjust schedules or alter specifications if there is a risk of going over budget.

The length of the progress report will depend on the complexity of the project. Dale Brandt's assessment of the progress his construction company is making in renovating Dr. Burke's clinic is given in a two-page letter in Figure 8.4 (pp. 249–250). A shorter report on organizing a workshop on time management, however, might require only a short e-mail.

Frequency of Progress Reports

Progress reports can be written daily, weekly, monthly, quarterly, or annually. Your specific job and your employer's needs will dictate how often you have to keep others informed of your progress. Contractor Brandt determined that three reports, spaced four to six weeks apart, would be necessary to keep Dr. Burke posted. Figure 8.4 is the second of those reports.

Parts of Progress Reports

Progress reports should contain information on (1) the work you have done, (2) the work you are currently doing, and (3) the work you will do.

How to Begin a Progress Report

In a brief introduction, cover the following:

- indicate why you are writing the report
- provide any necessary project titles or codes and specific dates
- help readers recall the job you are doing for them

If you are writing an initial progress report, supply background information in the opening. But, if you are submitting a subsequent progress report, your introduction should remind the reader about where your previous report left off and where the current one begins. Note how Dale Brandt's first paragraph in Figure 8.4 calls attention to the successful continuity of his work.

How to Continue a Progress Report

The body of the report should provide significant details about costs, materials, personnel, and times for the major stages of the project.

- Emphasize completed tasks, not false starts. If you report that the carpentry work or painting is finished, readers do not need an explanation of paint viscosity or geometrical patterns.
- Omit routine or well-known details ("I had to use the library when I wanted to read the back issues of *Safety News* that were not on the Internet").

The second of three progress reports from a contractor to a customer.   **Figure 8.4**

Brandt Construction Company

Halsted at Roosevelt, Chicago, Illinois 60608-0999 • 312-555-3700 • Fax: 312-555-1731
http://www.brandt.com

April 28, 2009

Dr. Pamela Burke
1439 Grand Avenue
Mount Prospect, IL 60045-1003

Dear Dr. Burke:

Here is my second progress report about the renovation work being done at your new clinic at Hacienda and Donohue. Work proceeded satisfactorily in April according to the plans you had approved in March.

Begins with key information: project status

Review of Work Completed in March

As I informed you in my first progress report on March 31, we tore down the walls, pulled the old wiring, and removed existing plumbing lines. All the gutting work was finished in March.

Recaps for background

Work Completed During April

By April 7, we had laid the new pipes and connected them to the main septic line. We also installed the two commodes, the four standard sinks, and the utility basin. The heating and air-conditioning ducts were installed by April 13. From April 16–20, we erected soundproof walls in the four examination rooms, the reception area, your office, and the laboratory. We had no problems reducing the size of the reception area by five feet to make the first examination room larger, as you had requested.

Summarizes current accomplishments

Problems with the Electrical System

We had difficulty with the electrical work, however. The number of outlets and the generator for the laboratory equipment required extra-duty power lines that had to be approved by both Con Edison and Cook County inspectors. The approval slowed us down by three days. Also, the vendor, Midtown Electric, failed to deliver the recessed lighting fixtures by April 27 as promised. Those fixtures and the generator are now being installed. Moreover, the cost of those fixtures will increase the material budget by **$5,288.00**. The cost for labor is as we had projected—**$89,450**.

Identifies problems, how they were solved, and costs involved

Work Remaining

The finishing work is scheduled for May. By May 11, the floors in the examination rooms, laboratory, washrooms, and hallways should be tiled and the reception area

Specifies what work remains and when it will be completed

Continued

Figure 8.4   (Continued)

page 2

Projects successful completion of work

Promises to keep reader informed

and your office carpeted. By May 13, the reception area and your office should be paneled and the rest of the walls painted. If everything stays on schedule, touch-up work is planned for May 14–18. You should be able to move into your new clinic by May 21.

You will receive a third and final progress report by May 14. Thank you again for your business and the confidence you have placed in our company.

Sincerely yours,

Dale Brandt

Dale Brandt

- Describe in the body of your report any snags you encountered that may affect the work in progress (see Dale Brandt's section on electrical problems in Figure 8.4). It is better for the reader to know about trouble early in the project so that appropriate changes or corrections can be made.

How to End a Progress Report

The conclusion should give a timetable for the completion of duties or submission of the next progress report. Give the date by which you expect work to be completed. Be realistic; do not promise to have a job done in less time than you know it will take. Readers will not expect miracles, only informed estimates. Even so, any conclusion must be tentative. Note that the good news Dale Brandt gives Dr. Burke about moving into her new clinic is qualified by the words "If everything stays on schedule." He is also well aware of the "you attitude" by thanking Dr. Burke again for her business.

▉ Travel/Trip Reports

Reporting on the trips you take is an important professional responsibility. In documenting what you did and saw, travel reports keep readers informed about your efforts and how they affect ongoing or future business. Travel reports also should be written after you attend a convention or sales meeting or call on customers.

Questions Travel Reports Answer

Specifically, travel reports should answer these questions for your readers.

- Where did you go?
- When did you go?
- Why did you go?
- Whom did you see?
- What did they tell you?
- What did you do about it?

For a business trip, you are also likely to have to inform readers how much it cost and to supply them with receipts for all of your expenses.

Common Types of Travel/Trip Reports

Travel reports can cover a wide range of activities and are called by different names to characterize those activities. Most likely, you will encounter the following three types of travel reports.

1. Site inspection reports. These reports inform managers about conditions at a branch office or plant, a customer's business, or on the advisability of relocating an office or other facility. After visiting the site, you will determine whether it meets your employer's (or customer's) needs. Site inspection reports provide information about the physical plant, the environment (air, soil, water, vegetation), or computer or financial operations.

Figure 8.5, which begins with a recommendation, is a report written to a district manager interested in acquiring a new site for a fast-food restaurant.

2. Field trip reports. These reports, often assigned in a course, are written after a visit to an office complex, hospital, detention center, or other facility to show what you have learned about the operation of a facility. You will be expected to describe how an institution is organized, the technical procedures and/or equipment it uses, pertinent ecological conditions, or the ratio of one group to another. The emphasis in such reports is on the educational value of the trip. For example, "From my visit to Water Valley, I learned a great deal about the health care delivery system at an extended care facility; this will help me during my internship next term."

3. Home health or social work visits. Nurses, social workers, and probation officers, for instance, report daily on their visits to patients and clients. Their reports describe clients' lifestyles, assess needs, and make recommendations based upon a variety of sources—clients, health care professionals, charitable organizations, and the like. These reports are often divided into Purpose of the Visit, Description of the Visit, and Action Taken as a Result of the Visit.

How to Gather Information for a Trip/Travel Report

Regardless of the kind of trip report you have to write, your assignment will be easier and your report better organized if you follow these suggestions.

1. Before you leave on the field trip, site inspection, or visit, be sure you are prepared as follows:

Figure 8.5   A site inspection report using a map.

VAIL'S

TO: Pretha Bandi   DATE: July 2, 2008
FROM: Beth Armando   *B.A.*   SUBJECT: New Site for Vail's #7
         Development Department

Begins with most important detail

Recommendation

The best location for the new Vail's Chicken House is the vacant Dairy World restaurant at the northeast corner of Smith and Fairfax Avenues—1701 Fairfax. I inspected this property on June 21 and 22 and also talked to Kim Shao, the broker at Crescent Realty representing the Dairy World Company.

Gives contact information

The Location

Please refer to the map below. Located at the intersection of the two busiest streets on the southeast side, the property will allow us to take advantage of the traffic flow to attract customers. Being only one block west of the Cloverleaf Mall should also help business.

Provides necessary background details

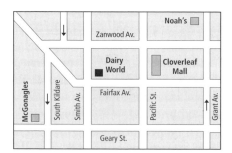

Includes map and traffic flow information essential for reader's purpose

Another benefit is that customers will have easy access to our location. They can enter or exit Dairy World from either Smith or Fairfax. Left turns on Smith are prohibited from 7 a.m. to 9 a.m., but since most of our business is done after 11 a.m., the restriction poses few problems.

Denver, CO 87123   (303) 555-7200   http://www.vails.com

Continued

(Continued)   **Figure 8.5**

Pretha Bandi
July 2, 2008
Page 2

Area Competition
Only two other fast-food establishments are in a one-mile vicinity. McGonagles, 1534 South Kildare, specializes in hamburgers; Noah's, 703 Zanwood, serves primarily seafood entrées. Their offerings will not directly compete with ours. The closest fast-food restaurant serving chicken is Johnson's, 1.8 miles away.

Assesses the location, in light of the competition

Parking Facilities
The parking lot has space for 45 cars, and the area at the south end of the property (38 feet $\times$ 37 feet) can accommodate 14–15 cars. The driveways and parking lot were paved with asphalt last March and appear to be in excellent condition. We will also be able to make use of the drive-up window on the north side of the building.

Gives only the most essential facts audience needs on parking, seating capacity, and alterations

The Building
The building has 3,993 square feet of heated and cooled space. The air-conditioning and heating units were installed within the last fifteen months and seem to be in good working order; nine more months of transferable warranty remain on these units.

The only major changes we must make are in the kitchen. To prepare items on the Vail's menu, we need to add three more exhaust fans (there is only one now) and expand the grill and cooking areas. The kitchen also has three relatively new sinks and offers ample storage space in the sixteen cabinets.

Writing is clear and concise

The restaurant has a seating capacity of up to 54 persons; 10 booths are covered with red vinyl and are comfortably padded. A color-coordinated serving counter could seat 8 to 10 patrons. The floor does not need to be retiled, but the walls will have to be painted to match Vail's color decor.

Does not overwhelm with petty details

    a. Obtain all necessary names; street, e-mail, and website addresses; and relevant telephone, cell, and fax numbers.

    b. Check files for previous correspondence, case studies, or terms of contracts or agreements.

    c. Download work orders, instructions, or other documents pertinent to your visit, for example, websites and ads.

    d. Bring a PDA or notebook with you.

    e. Locate a map of the area and get the directions you'll need beforehand. (Both maps and directions can be obtained easily at *http://www.mapquest.com*, *http://maps.yahoo.com*, and *http://maps.google.com*).

    f. Bring a record of appointment times and locations as well as the job titles of the people whom you expect to meet.

    g. Bring a camcorder, tape recorder, camera, or calculator, if necessary, to record important data.

2. When you return from your trip, keep the following hints in mind as you compile your report:

    a. Write your report promptly. If you put it off, you may forget important items.

    b. When a trip takes you to two or more widely separated places, note in your report when you arrived at each place and how long you stayed.

    c. Exclude irrelevant details, such as whether the trip was enjoyable, what you ate, or how delighted you were to meet people.

    d. Check to make sure you have listed names and calculated figures correctly. Mistakes in math make you look bad.

Incident Reports

The short reports discussed thus far in this chapter have dealt with routine work. They have described events that were anticipated, planned, or supervised. But every business or agency runs into unexpected trouble that delays routine work. More often than not, these circumstances need to be documented in an incident report. Employers and, on some occasions, government inspectors, insurance agents, and attorneys must be informed about those events that interfere with or threaten normal, safe operations. Incident reports can be submitted as a memo, as in Figure 8.6 (pp. 255–256), or on specially prepared forms your employer or government agency expects you to follow.

When to Submit an Incident Report

An incident report is required when there is, for example:

- an accident—fire, automobile, physical injury
- a law enforcement offense
- an environmental danger, including a computer virus
- a machine breakdown
- a delivery delay
- a cost overrun
- a production slowdown

An incident report in memo format.   **Figure 8.6**

THE GREAT HARVESTER RAILROAD
Des Moines, IA 50306-4005
http://www.ghrr.com

TO: Angela O'Brien, District Manager
            James Hwang, Safety Instructor
FROM: Nick Roane, Engineer *Nick Roane*
DATE: March 5, 2008
SUBJECT: Derailment of Train 26 on March 5, 2008

TYPE OF INCIDENT
Two grain cars went off the track while I was driving Engine 457. There were no injuries to the crew.

Begins with most impor- tant details

DESCRIPTION OF INCIDENT
At 7:20 a.m. on March 5, 2008, I was traveling north at a speed of 52 miles an hour on the single main-line track four miles east of Ridgeville, Illinois. Weather conditions and visibility were excellent. Suddenly the last two grain cars, 3022 and 3053, jumped the track. The train automatically went into emergency braking and stopped immediately. But the train did not stop before both grain cars turned at a 45° angle. After checking the cars, I found that half the contents of their loads had spilled. The train was not carrying any hazardous chemical shipments.

Gives precise time, location

Describes what happened

I notified Supervisor Bill Purvis at 7:40 a.m., and within 45 minutes he and a section crew arrived at the scene with rerailing equipment. The section crew removed the two grain cars from the track, put in new ties, and made the main-line track passable by 9:25 a.m. At 9:45 a.m. a vacuum car arrived with Engine 372 from Hazlehurst, Illinois, and its crew proceeded with the cleanup operation. By 10:25 a.m. all the spilled grain was loaded onto the cars brought by the Hazlehurst train. Bill Purvis notified Barnwell Granary that their shipment would be at least three hours late.

Explains what was done

CAUSES OF INCIDENT
Supervisor Purvis and I checked the stretch of train track where the cars derailed and found it to be heavily worn. We believe that a fisher joint slipped when the grain cars hit it, and the track broke. You can see the location of the cracked fisher joint in the graphic below.

Determines likely cause

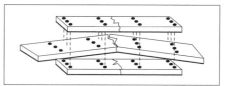

Supplies ex- ploded visual

Continued

Figure 8.6   (Continued)

Offers precise recommendations to solve or prevent problem from recurring

Page 2

RECOMMENDATIONS
We made the following recommendations to the switch yard in Hazlehurst to be carried out immediately.
1. Check the section of track for ten miles on either side of Ridgeville for any signs of defective fisher joints.
2. Repair any defective joints at once.
3. Instruct all engineers to slow down to five to ten mph over this section of the track until the rail check is completed.

The incident report above on the train accident was submitted as a memo by the engineer.

Parts of an Incident Report

Include the following information in your incident report. Because it can contain legally sensitive information, for which the reader needs a hard copy and paper trail, an incident report should not be sent as an e-mail. Note how Figure 8.6 includes detailed and accurate information on these parts.

1. Identification details.   Indicate who was involved. List employee identification numbers. Record titles, department, and employment identification. For customers or victims, record home addresses, phone numbers, and places of employment. Insurance companies will also require policy numbers. Be sure to include any damaged equipment model/serial numbers.

2. Type of incident.   Briefly identify the incident—personal injury, fire, burglary, equipment failure. Identify any part(s) of the body precisely. "Eye injury" is not enough; "injury to the right eye, causing bleeding" is better.

3. Time and location of the incident.   Include precise date (not "Thursday") and time (a.m. or p.m.).

4. Description of what happened.   Let readers know exactly what happened and why, how it occurred, and what led up to the incident.

5. What was done after the incident.   Describe the action you took to correct conditions, to get things back to normal, what was done to treat the injured, to make the environment safer, to speed a delivery, to repair damaged equipment.

6. What caused the incident.   Make sure your explanation is consistent with your description of what happened. Pinpoint the trouble. In Figure 8.6, for example, the defective fisher joint is listed under the heading "Causes of Incident."

7. Recommendations.   Recommendations about preventing the problem from recurring may involve repairing any broken parts, as in Figure 8.6, calling a special safety meeting, asking for further training, adapting existing equipment, doing emergency planning, or modifying schedules.

Protecting Yourself Legally

An incident report is an official legal record that can be used by law enforcement and attorneys in court to establish negligence and liability on your and your company's part. An incident report frequently concerns the two topics over which powerful legal battles are waged—health and property.

You have to be very careful about collecting and recording details. Make sure your report is not sketchy or incomplete. To avoid these errors, you may have to interview employees or bystanders; travel to the incident site; check manuals, code books, or other guides; consult safety experts; or research records/archives.

To ensure that what you write is legally proper, follow these guidelines.

1. Submit your report promptly and sign or initial it.   Any delay might be seen as a cover-up. Send your report to the appropriate parties immediately after you have gathered the necessary information.

2. Be accurate, objective, and complete.   Recount clearly what happened in the order it took place. Never omit or distort facts; the information may surface later, and you could be accused of a cover-up. Do not just write "I do not know" for an answer. If you are not sure, state why. Also be careful that there are no discrepancies in your report.

3. Give facts, not opinions.   Provide a factual account of what actually happened, not an emotional, biased interpretation of events or one based on speculation or hearsay. Vague words such as "I guess," "I wonder," "apparently," "perhaps," or "possibly" weaken your objectivity. Identify witnesses or victims by giving names, addresses, places of employment, and so on. Stick to details you witnessed. Keep in mind that stating what someone else saw is regarded as hearsay and therefore is not admissible in a court of law. State only what *you* saw or heard. When you describe what happened, avoid drawing uncalled-for conclusions. Consider the following statements of opinion and fact:

Opinion: The patient seemed confused and caught himself in his IV tubing.
    Fact: The patient caught himself in his IV tubing.

Opinion: The equipment was defective.
    Fact: The bolt was loose.

Be careful, too, about blaming someone. Statements such as "Baxter was incompetent" or "The company knew of the problem but did nothing about it" are libelous remarks.

In law enforcement work, further identify suspects by their aliases and by any distinctive characteristics—for example, "jagged 4-inch scar on left forearm."

4. Do not exceed your professional responsibilities. Answer only those questions you are qualified to answer. Do not presume to speak as a detective, an inspector, a physician, or a supervisor. Do not represent yourself as an attorney or a claims adjuster in writing the report. Finally, don't act like a judge and jury by placing blame or taking one side over another.

Writing Successful Proposals

A proposal is a detailed plan of action that a writer submits to a reader or group of readers for approval. The readers are usually in a position of authority—supervisors, managers, department heads, company buyers, boards of private foundations, elected officials, military or civic leaders—to endorse or reject the writer's plan.

Proposals are written for many purposes and many different audiences. You can write an internal proposal, for example to your boss, seeking authorization to hire staff, change a procedure, or to purchase a new piece of equipment for the office. Or, you can write a sales proposal to potential customers, offering a product or a service (such as offering to supply a fire chief with special firefighting gear, or offering an office manager a line of ergonomically designed furniture).

Depending on the job, proposals can vary greatly in size and in scope. A proposal to your employer could easily be conveyed in a few pages, the length of a short report. To propose doing a small job for a prospective client—redecorating a waiting room in an accountant's office—a letter with information on costs, materials, and a timetable might suffice. The sales letter in Figure 4.8 (p. 100) illustrates a short proposal in letter format. Proposals can be *unsolicited*—that is, they originate with you—or they can be *solicited,* requested by a company or organization, as in Figure 8.9 (pp. 269–271).

Proposals Are Persuasive Plans

Proposals, whether large or small, must be highly persuasive to succeed. Without your audience's approval, your plan will never go into effect, however accurate and important you think it is. Your enthusiasm is not enough to persuade readers; you have to supply hard evidence. Your proposals must convince readers that your plan will help them improve their businesses, make their jobs easier, save them money, enhance their image, or all of these.

Every proposal you write must exhibit a "can do" attitude, putting the reader and his or her company's needs at the center of your work. Show readers how approving your plan will save them time and money or will improve employee morale or customer satisfaction. The tone of your proposal should be **"Here is what I can do for you."** Yates Engineering has won millions of dollars of business through its reader-centered proposals. Its slogan is "On time . . . within budget . . . to your satisfaction." Time, budget, and your readers' satisfaction and convenience are among the key ingredients of a winning proposal. Notice how the Cingular advertisement in Figure 8.7 below assures customers that they will receive global access, technical expertise, and high-quality service.

An example of a "can-do" attitude. **Figure 8.7**

Reprinted courtesy of Cingular.

Proposals Frequently Are Collaborative Efforts

Like many other business and technical writing, proposals often are the product of teamwork. Even a short in-house proposal is often researched and put together by more than one individual in the company or agency.

Many times, individual employees will pull together information from their separate areas (such as graphics and design, finance, marketing, technology, transportation, and even legal) and put it into a proposal that each member of the team then reads and revises until the team agrees that the document is ready to be released.

Guidelines for Writing a Successful Proposal

The following guidelines will help you to persuade your audience to approve your plan. Refer to these guidelines and Figures 8.8 (pp. 262–265) and 8.9 (pp. 269–270) both before and while you formulate your plan.

1. Approach writing a proposal as a problem-solving activity.   Your purpose should reflect your ability to solve problems. Convince your audience you know what their needs are and that you will meet their needs.

2. Regard your audience as skeptical.   Even though you offer a proposal that you think will benefit your readers, do not be overconfident that they will automatically accept it. To determine whether your proposal is feasible, readers will study it carefully. If your proposal contains errors and inconsistencies, omits important information, or deviates from what your readers are looking for, then your audience is likely to reject it.

3. Research your proposal topic thoroughly.   A winning proposal is *not* based on a few well-meaning, general suggestions. To provide the detailed information necessary and to convince readers that you know your proposal topic inside and out, you will have to do your homework. Research your topic by shopping for the best prices, comparing prices and services with what the competition offers, verifying schedules, visiting customers, making site visits, and interviewing individuals.

4. Scout out what your competitors are doing.   Become familiar with your competitors' products or services, have a fair idea about their market costs, and be able to show how your company's work is better overall. Read their websites and print publications (such as catalogs and marketing brochures) very carefully. Let readers know you have done your homework.

5. Prove that your proposal is workable.   What you propose should be consistent with the organization and capabilities of the company. For instance, recommending that a small company with fifty employees triple its work force to implement your plan would be foolish and risky. Your proposal should not contain statements such as "Let's see what happens."

6. Be sure your proposal is financially realistic.   "Is it worth the money?" is a bottom-line question you can expect from your readers. For example, recommending that your company spend $20,000 to solve a $2,000 problem is just not feasible. Above all, convince your readers that the benefits are worth the costs. Note how Figure 8.8 (pp. 262–265) saves the bank money and Figure 8.9 (pp. 269–271) stresses that the costs are under control.

7. Package your proposal attractively.   Make sure that your proposal is well presented: inviting, attractive, and easy to read, and that all visuals are clear and appropriately placed. The visual appearance of your proposal can contribute greatly to whether it is accepted.

■ Internal Proposals

The primary purpose of an internal proposal, such as the one shown in Figure 8.8, is to offer a realistic and constructive plan to help your company run its business more efficiently and economically.

On your job you may discover a better way of doing something or a more efficient way to correct a problem. You believe that your proposed change will save your employer time, money, or further trouble. (Note how Tina Escobar and Oliver Jabur identified and researched a more effective and less costly way for Community Federal Bank to conduct business and to satisfy its customers in Figure 8.8, pp. 262–265.)

Generally speaking, your proposal will be an informal, in-house message, so a brief (usually one-to-three page) memo or e-mail should be appropriate. You decide to notify your department head, manager, or supervisor, or your employer may ask you for specific suggestions to solve a problem that has already been identified.

Typical Topics for Internal Proposals

An internal proposal can be written about a variety of topics, including the following:

- purchasing new or more advanced equipment to replace obsolete or inefficient computers, transducers, robots, and the like, or upgrading equipment
- obtaining document security software and offering training sessions to show employees how to use it
- recruiting new employees or retraining current ones on a new technique or process
- eliminating a dangerous condition or reducing an environmental risk to prevent accidents—for employees, customers, or the community at large
- improving technology/communication within or between departments of a company or agency
- expanding work space or making it greener, more private, ergonomically beneficial to employees, or more inviting to customers

As the list shows, internal proposals cover almost every activity or policy that can affect the day-to-day operation of a company or agency.

Figure 8.8   An internal unsolicited proposal.

**COMMUNITY
FEDERAL BANK**

http://www.comfedbank.com

Equal Housing
Lender

Powell
617-584-5200

Monroe
781-413-6000

Langston
508-796-3009

TO: Michael L. Sappington, Executive Vice President
Dorothy Woo, Langston Regional Manager

FROM: Tina Escobar, Oliver Jabur, ATM Services

DATE: June 11, 2008

RE: A proposal to install an additional ATM at the Mayfield Park
branch

*Clearly states
why proposal is
being sent*

PURPOSE

We propose a cost-effective solution to what is a growing problem
at the Mayfield Park branch in Langston: inefficient servicing of
customer needs and rising personnel costs. We recommend that
you approve the purchase and installation, within the next two to
three months, of another ATM at Mayfield. Such action is consistent
with Community's goals of expanding branch banking services
and promoting our image as a self-serve yet customer-oriented
institution.

*Identifies
problem by
giving reader
necessary
background
information
based on
research*

THE PROBLEM WITH CURRENT SERVICES AT MAYFIELD PARK

Currently, we employ four tellers at Mayfield. Too much is being spent on
personnel/salary for routine customer transactions. In fact, as determined
by teller activity reports, nearly 25 percent of the four tellers' time each
week is devoted to routine activities easily accommodated by the installation of another ATM. Outlined in the table below is a breakdown of teller
activity for the month of May:

*Provides easy-
to-read table*

Teller #	Total Transactions	Routine Transactions
1	6,205	1,551
2	5,989	1,383
3	6,345	1,522
4	6,072	1,518
	24,611	5,974

*Divides prob-
lem into
parts—volume,
financial, per-
sonnel, cus-
tomer service*

Clearly, we are not fully using our tellers' sales abilities when they are
kept busy with routine activities. To compound the problem, we expect
business to increase by at least 25 percent at Mayfield in the next few
months, as projected by this year's market survey. If we do not install

Continued

page 2

an additional ATM, we will need to hire a fifth teller, at an annual cost of $20,800 ($15,500 base pay plus approximately 30 percent for fringes), for the additional 6,000 transactions we project.

Most important, customer needs are not being met efficiently at Mayfield. Recent surveys done for Community Federal by Watson-Perry demonstrate that our customers are inconvenienced by not having one more ATM at Mayfield. They are unhappy about long waits both at the ATMs and at the teller windows to do simple banking business, such as deposits, withdrawals, and loan payments. Conversations we had with manager Rachael Harris-Koyoto at the Mayfield branch confirm customers' complaints.

Verifies that problem is widespread

Ultimately, the lack of another ATM at Mayfield Park hurts Community's image. With plentiful ATMs available to Mayfield residents at local stores and at other banks, our institution risks having customers and potential customers go elsewhere for their banking needs. We not only miss the opportunity of selling them our other services but also risk losing their business entirely.

Emphasizes possible future problems

A SOLUTION TO THE PROBLEM

Purchasing and installing an additional ATM at Mayfield Park will result in significant savings in personnel costs and time. We will

- Save money by not having to hire a fifth teller
- Allocate teller duties more efficiently and productively by assisting customers with questions and transactions not handled through an ATM, such as opening a new account; purchasing savings bonds, CDs, traveler's checks and foreign currency; and doing Internet banking.
- Increase time for tellers to cross-sell our services, including our line of nontraditional banking products—annuities, mutual funds, debit cards, and global market accounts
- Service customer retirement options by having tellers track IRAs, 401(k)s, 403(b)s, and Simples
- Improve customer satisfaction by giving them the option of meeting their banking needs electronically or through a teller
- Ease the stress on tellers at Mayfield Park

Relates solution to individual parts of the problem

It is feasible to install another ATM at Mayfield. This location does not pose the difficulties as at some older branches. Mayfield offers ample room to install a drive-up ATM in the stubbed-out fourth drive-up lane. It is away from the heavily congested area in front of the bank, yet it is

Shows problem can be solved and stresses how

Continued

Figure 8.8   (Continued)

easily accessible from the main driveway and the side drive facing Commonwealth Avenue, as the photograph below shows.

Photo shows location has room for additional ATM

Photo courtesy Taylor Wilson

Documents that work can be done on time and highlights advantage of doing it now

Judging from our previous experience, the ATM could be installed and operational within one to two months. Moreover, by authorizing the expenditure at Mayfield within the next month, you will ensure that additional ATM service is available long before the busy Christmas season.

COSTS

The costs of implementing our proposal are as follows:

Itemizes costs

Diebold drive-up ATM	$28,000.00
Installation fee	2,000.00
Maintenance (1 year)	1,500.00
	$31,500.00

Interprets costs for reader

This $31,500, however, does not truly reflect our annual costs. We would be able to amortize, for tax purposes, the cost of installation of the ATM over five years. Our annual expenses would, therefore, look more like this:

$30,000 (28,000 + 2,000) divided by 5 years =
$6,000 + 1,500 (maintenance), or $7,500 per year.

Continued

(Continued) **Figure 8.8**

---

page 4

Compared with the $20,800 a year the bank would have to expend for a fifth teller position at Mayfield, the annual depreciated cost for the ATM ($7,500) reduces by nearly two-thirds the amount of money the bank will have to spend for much more efficient customer service.

CONCLUSION

Authorizing another ATM for the Mayfield Park branch is both feasible and cost effective. Endorsement of this proposal will save our bank more than $13,000 in teller services annually, reduce customer complaints, and increase customer satisfaction and approval. We will be happy to discuss this proposal with you anytime at your convenience, and answer any questions you may have.

Proves change is cost-effective

Stresses benefits for reader and bank as a whole

Following the Proper Chain of Command

Writing an internal proposal requires you to be aware of and sensitive to office politics. It may be wise first to meet with your boss to see if she or he has already identified the problem or has specific suggestions on how to solve it. Then you might provide your boss with a draft and ask for revisions or feedback.

You cannot assume that your reader(s) will automatically agree with you that there is a problem or that your plan is the only way to tackle it. To be successful, write your internal proposal keeping in mind the needs and likes of your boss and others who may have to sign off on it. Remember that your boss will expect you to be very convincing about both the problem you say exists and the changes you are advocating in the workplace under his or her supervision. Don't step on corporate toes. Similarly, you may need the approval of individuals in other offices, departments, or branches of your organization.

Ethically Identifying and Resolving Readers' Problems

When you prepare an internal proposal, you need to be aware of the ethical obligations you have and the ways to meet them. Here are three important guidelines:

1. Consider the implications of your plan company-wide. The change you propose (transfers, new schedules or technology, new hires) may have sweeping and potentially disruptive implications in another division of your company.
2. Keep in mind what impact your change may have for co-workers from cultural traditions other than your own. In addition to speaking with your boss, it would be wise to consult your human resources director.

3. Never submit an internal proposal that offers an idea that you think will work, but which relies on someone else to supply the specific details, the plan, on how and when it will work.

Organization of Internal Proposals

A short internal proposal follows a relatively straightforward plan of organization, from identifying the problem to solving it. Internal proposals usually contain four parts, as shown in Figure 8.8: **purpose, problem, solution,** and **conclusion.** Refer to the figure (pp. 262–265) as you read the following discussion.

The Purpose

Begin your proposal with a brief statement of why you are writing to your supervisor: "I propose that . . ." State right away why you think a specific change is necessary now. Then succinctly define the problem and emphasize that your plan, if approved by the reader, will solve that problem.

The Problem

Prove that a problem exists. Document its importance for your boss and your company; as a matter of fact, the more you show, with concrete evidence, how the problem affects the boss's work (and area of supervision), the more likely you are to persuade him or her to act. Here are some guidelines for documenting a problem.

- Avoid vague (and unsupported) generalizations such as "We're losing money each day with this procedure (piece of equipment)"; "Costs continue to escalate"; "The trouble occurs frequently in a number of places"; "Numerous complaints have come in"; "If something isn't done soon, more problems will result."
- Provide quantifiable details about the problem, such as the amount of money or time a company is actually losing per day, week, or month. Document the financial trouble so that you can show in the next section how your plan offers an efficient and workable solution.
- Indicate how many employees (or work hours) are involved or how many customers are inconvenienced or endangered by a procedure or condition. Notice how Escobar and Jabur include such information in a table in their proposal in Figure 8.8.
- Verify how widespread a problem is or how frequently it occurs by citing specific occasions. Again, see how Escobar and Jabur cite evidence from the Watson-Perry survey and the interviews they conducted with the manager of the Mayfield branch.
- Relate the problem to an organization's image, corporate reputation, or influence (where appropriate). Pinpoint exactly how and where the problem lessens your company's effectiveness or hurts its standing in the market. Indicate who is affected and how the problem affects your company's business community, as the writers do especially well in the first paragraph of Figure 8.8 (p. 262).

The Solution or Plan

In this section describe how you are to make the change you propose and want approved and demonstrate the likelihood of its succcess. Your reader will again expect to find factual evidence. Be specific. Supply details that answer the following questions: (1) Is the plan workable—can it be accomplished here in our office or plant? (2) Is it cost-effective—will it really save us money in the long run and not lead to even greater expenses?

To get the reader to say "Yes" to both questions, supply the facts you have gathered as a result of your research. For example, if you propose that your firm buy a new piece of equipment, do the necessary homework to locate the most efficient and cost-effective model available, as Tina Escobar and Oliver Jabur did for their proposal in Figure 8.8.

- Supply the dealer's name, the costs, major conditions of service and training contracts, and warranties.
- Describe how your firm could use the equipment to obtain better or quicker results in the future.
- Document specific tasks the new equipment can perform more efficiently at a lower cost than the equipment now in use.

A **proposal to change a procedure** must address the following questions.

- How does the new (or revised) procedure work?
- How many employees or customers will be affected by it?
- When will it go into operation?
- How much will it cost the employer to change procedures?
- What delays or losses in business might be expected while the company switches from one procedure to another?
- What employees, equipment, or locations are already available to accomplish the change?

The costs, in fact, will be of utmost importance. Make sure you supply a careful and accurate budget. Moreover, make the costs attractive by emphasizing how inexpensive they are compared to the cost of *not* making the change, as Escobar and Jabur do in the section labeled **"Costs."** Be sure to double-check your math.

It is also wise to raise alternative solutions, before the reader does, and to discuss their disadvantages. Notice how Tina Escobar and Oliver Jabur do that in Figure 8.8 by showing why installing an ATM is more feasible than hiring a fifth teller.

The Conclusion

Your conclusion should be short—a paragraph or two at the most. Remind readers that (a) the problem is ongoing and serious, (b) the reason for change is justified and will be beneficial to your organization, and (c) action needs to be taken. Re-emphasize the most important benefits. Escobar and Jabur stress the savings that the bank will see by following their plan as well as the increase in customer satisfaction. Also indicate that you are willing to discuss your plan with the reader and seek his/her input, a necessity in arguing for a corporate change at any level.

■ Sales Proposals

A sales proposal is the most common type of proposal. Its purpose is to sell your company's products or services for a set fee. A short sales proposal is a marketing tool that includes a sales pitch as well as a detailed description of the work you propose to do. Figure 8.9 (pp. 269–271) contains a sales proposal in response to a company's request.

The Audience and Its Needs

Your audience will usually be one or more executives who have the power to approve or reject a proposal. Unlike readers of an internal proposal, your audience for a sales proposal may be even more skeptical since they may not know you or your work. You can increase your chances of success by trying to anticipate their questions such as:

- Does the writer's firm understand our problem?
- Can the writer's firm deliver the services it promises?
- Can the job be completed on time?
- Is the budget reasonable and realistic (neither inflated nor too low)?
- What assurances does the writer offer that the job will be done exactly as proposed?
- How has the writer demonstrated his/her trustworthiness?

Answer each of these questions by demonstrating how your product or service is tailored to the customer's needs.

Make sure, too, that your proposal has a competitive edge. Readers will compare your plan with those they receive from other proposal writers. Your proposal has to convince readers that the product and the service your company offers are more reliable, economical, efficient, and timely than those of another company. Here is where your homework pays off.

Organizing Sales Proposals

Most sales proposals include the following elements: introduction, description of the proposed product or service, timetable, costs, qualifications of your company, and conclusion.

Introduction

The introduction to your sales proposal can be a single paragraph in a short sales proposal or several pages in a more complex one. Basically, your introduction should prepare readers for everything that follows in your proposal. The introduction itself may contain the following sections, which sometimes may be combined.

1. Statement of purpose and subject of proposal. Tell readers why you are writing and identify the specific subject of your work. Briefly define the solution you propose. Tell them exactly what you propose to do for them. Be clear about what your plan covers and, if there could be any doubt, what it does not do.

A proposal in response to a request from a company.   **Figure 8.9**

Reynolds Interiors • 250 Commerce Avenue S.W. • Portland, OR 97204-2129

January 21, 2009

Mr. Floyd Tompkins, Manager
General Purpose Appliances
Highway 11 South
Portland, OR 97222

Dear Mr. Tompkins:

In response to your request listed on your website for bids for an
appropriate floor covering at your new showroom, Reynolds Interiors
is pleased to submit the following proposal. We appreciated the
opportunity to visit your facility in order to submit this proposal.

After carefully reviewing your specifications for a floor covering and
inspecting your new facility, we believe that **Armstrong Classic
Corlon 900** is the most suitable choice. We are enclosing a sample of
the Corlon 900 so you can see how carefully it is constructed.

Corlon's Advantages
Guaranteed against defects for a full three years, **Corlon** is one of the
finest and most durable floor coverings manufactured by Armstrong.
It is a heavy-duty commercial floor 0.085-inch thick for protection.
Twenty-five percent of the material consists of interface backing;
the other 75 percent is an inlaid wear layer that offers exceptionally
high resistance to everyday traffic. Traffic tests conducted by the
Independent Floor Covering Institute have repeatedly proved the
superiority of **Corlon's** construction and resistance.

Another important feature of **Corlon** is the size of its rolls. Unlike
other leading brands of similar commercial flooring—Remington or
Treadmaster—**Corlon** comes in 12-foot-wide rather than

http://www.reynolds.com • 503-555-8733 • Fax: 503-555-1629

*Begins with a
reference to
company's
request for bids*

*Identifies best
solution*

*Describes prod-
uct's features
that will bene-
fit reader*

*Distinguishes
product from
competitors'*

Continued

Figure 8.9   (Continued)

2

6-foot-wide rolls. This extra width will significantly reduce the number of seams on your floor, thus increasing its attractiveness and reducing the dangers of splitting.

Installation Procedures

The **Classic Corlon** requires that we use the inlaid seaming process, a technical procedure requiring the services of a trained floor mechanic. Herman Goshen, our floor mechanic, has more than fifteen years of experience working with the inlaid seam process. His professional work and keen sense of layout and design have been consistently praised by our customers.

Explains how job is done

Installation Schedule

We can install the **Classic Corlon** on your showroom floor during the first week of March, which fits the timetable specified in your request. The material will take three and one-half days to install and will be ready to walk on immediately. We recommend, though, that you not move equipment onto the floor for 24 hours after installation.

Gives realistic timetable

Costs

The following costs include the **Classic Corlon** tile, labor, and tax:

750 sq. yards of **Classic Corlon** at $23.50/sq. yd.	$ 17,625.00
Labor (28 hrs @ $18.00/hr.)	$ 504.00
Sealing fluid (10 gals. @ $15.00/gal.)	$ 150.00
Total	$ 18,279.00
Tax (5 percent)	$ 913.95
GRAND TOTAL	**$ 19,192.95**

Itemizes all costs

Our costs are $250.00 under those you specified in your request.

Continued

(Continued)   **Figure 8.9**

3

Reynolds's Qualifications

Reynolds Interiors has been in business for more than 28 years. In that time, we have installed many commercial floors in Portland and its suburbs. In the last year, we have served more than 60 customers, including the new multipurpose Tech Mart plant in downtown Portland. We would be happy to furnish you with a list of our satisfied customers.

Establishes history of service

Conclusion

Thank you for the opportunity to submit this proposal. We believe you will be pleased with the appearance and durability of an Armstrong **Corlon** floor. If we can provide you with any further information, please call us or visit us at our website.

Encourages reader to accept

Sincerely yours,

Neelow Singh

Neelow Singh
Sales Manager

Jack Rosen

Jack Rosen
Installation Supervisor

---

2. Background of the problem you propose to solve. Show readers that you are familiar with their problem and why it is important. In a solicited proposal like the one in Figure 8.9, this section is usually unnecessary because the potential client has already identified the problem and wants to know how you would address it. In that case, just point out how your company would solve the problem, mentioning your superiority over your competitors (see the fourth paragraph of the figure (p. 263)).

In an unsolicited proposal, you need to describe the problem in convincing detail, identifying the specific trouble areas. Depending on the type of proposal you submit, you may want to focus briefly on the dimensions of the problem—when it was first observed, who/what it most acutely affects, and the specific organizational/community context in which the problem is most troubling.

Description of the Proposed Product or Service

This section is the heart of your proposal. Before spending their money, customers will demand hard, factual evidence of what you claim can and should be done. Here are some points that your proposal should cover.

1. Carefully show potential customers that your product or service is right for them. Stress particular benefits of your product or service most relevant to your reader. Blend sales talk with descriptions of hardware.

2. Describe your work in suitable detail. Specify what the product looks like; what it does; and how consistently and well it will perform in the readers' office, plant, hospital, or agency. You might include a brochure; picture; diagram; or, as the writer of the proposal in Figure 8.9 does, a sample of your product for customers to study.

3. Stress any special features, maintenance advantages, warranties, or service benefits. Convince readers that your product is the most up-to-date and efficient one they could select. Highlight features that show the quality, consistency, or security of your work. For a service, emphasize the procedures you use, the terms of the service, even the kinds of tools you use, especially any state-of-the-art equipment.

Timetable

A carefully planned timetable shows readers that you know your job and that you can accomplish it in the right amount of time. Your dates should match any listed in a company's proposal request. Provide specific dates to indicate

- when the work will begin
- how the work will be divided into phases or stages
- when you will be finished
- whether any follow-up visits or services are involved

For proposals offering a service, specify how many times—an hour, a week, a month—customers can expect to receive your help; for example, spraying three times a month if your company offers exterminating service.

Costs

Make your budget accurate, complete, and convincing. But give customers more than merely the bottom-line cost. Show exactly what readers are getting for their money so that they can determine if everything they need is included. Itemize costs for

- specific services
- equipment and materials
- labor (by the hour or by the job)
- transportation
- travel
- training

If something is not included or is considered optional, say so—additional hours of training, replacement of parts, upgrades, and the like. If you anticipate a price increase, let the customer know how long current prices will stay in effect. That information may spur them to act favorably now.

Qualifications of Your Company

Emphasize your company's accomplishments and expertise in providing similar services and/or equipment. Mention the names of a few local firms for whom you have worked that would be able to recommend you. But never misrepresent your qualifications or those of the individuals who work with or for you. Your prospective client can verify whether you have in fact worked on similar jobs during the last five to six years.

Conclusion of Proposal

This is the "call to action" section of your sales proposal. Encourage your reader to approve your plan by stressing major benefits of your plan. Offer to answer any questions the reader may have. Some proposals end by asking readers to sign and return a copy of the proposal indicating their acceptance, as the proposal in Figure 8.9 does.

 Revision Checklist

Short Reports

☐ Had a clear sense of how my readers will use my short report.

☐ Provided significant, relevant information about costs, materials, personnel, and times so readers will know that my work consists of facts, not impressions.

☐ Double-checked all data—names, costs, figures, dates, places, and equipment numbers.

☐ Kept report concise and to the point.

☐ Used headings wherever feasible to organize and categorize information.

☐ Began report with statement of purpose that clearly described the scope and significance of my work.

☐ Incorporated tables and other pertinent visuals to display data whenever appropriate.

☐ Explained clearly what the data means.

☐ Determined that recommendations logically follow from the data and are realistic.

☐ Adhered to all ethical and legal requirements in writing an incident report.

Proposals

☐ Identified a realistic problem in my proposal—one that is restricted and relevant to my topic and my audience's needs.

☐ Tried effectively to convince audience that the problem exists and needs to be solved; incorporated the "you attitude" throughout.

☐ Incorporated details demonstrating the scope and importance of the problem.

Continued

Continued

☐ Persuasively emphasized benefits of solving the problem according to the proposal.

☐ Offered a solution that can be realistically implemented—that is, it is both appropriate and feasible for audience.

☐ Used specific figures and concrete details to show how proposal will save time and money.

☐ *For internal proposals:* Demonstrated how proposal benefits my company and my supervisor; followed the chain of command by discussing proposal with co-workers and/or supervisors who may be affected.

☐ *For sales proposals:* Related my product or service to prospective customer's needs; showed a clear understanding of those needs.

☐ Prepared a comprehensive and realistic budget; accounted for all expenses; itemized costs of products and services.

☐ Linked costs to benefits.

☐ Provided a timetable with exact dates for implementing proposal.

Exercises

Additional Activities related to writing short reports and proposals are located at **college.cengage .com/pic/ kolinconcise2e.**

1. Assume that you are a manager of a large apartment complex (200 units). Write a periodic report based on the following information—26 units are vacant, 38 soon will be vacant, and 27 soon will be leased (by June 1). Also add a section of recommendations to your supervisor (the head of the real-estate management company for which you work) on how vacant apartments might be leased more quickly and perhaps at increased rents. Consider important information such as decorating, advertising, installing a new security system, providing Internet access.

2. Assume you work for a household appliance store. Prepare a sales report based on the information contained in the following table. Include a recommendations section for your manager.

	Number Sold	
Product	*October*	*November*
Kitchen Appliances		
Refrigerators	72	103
Dishwashers	27	14
Freezers	10	36
Electric Ranges	26	26

Gas Ranges	10	3
Microwave Ovens	31	46
Laundry Appliances		
Washers	50	75
Dryers	24	36
Air Treatment		
Room Air Conditioners	41	69
Dehumidifiers	7	2

3. Submit a progress report to your writing teacher on what you have learned in his or her course so far this term, specifying which formatting and writing skills you want to develop in greater detail, and how you propose doing so. Mention specific memos, e-mails, letters, instructions, reports, or proposals you have written or will soon write.

4. Compose a site inspection report on any part of the college campus or plant, office, or store in which you work that might need remodeling, expansion, re-wiring for computer use, or new or additional air-conditioning or heating work.

5. Write a report to an instructor in your major or to an employer about a trip you have taken recently—to a museum, laboratory, health care agency, correctional facility, radio or television station, plant or factory, or office. Indicate why you took the trip, name the individuals you met on the trip, and stress what you learned and how that information will help you in course work or on your job. Include a relevant visual as part of your report.

6. Write an incident report about one of the following problems. Assume that it has happened to you. Supply relevant details and visuals in your report. Identify the audience for whom you are writing and the agency you are representing or trying to reach.
   a. After hydroplaning, your company car hits a tree and has a damaged front fender.
   b. You have been the victim of an electrical shock because an electrical tool was not grounded.
   c. You twist your back lifting a bulky package in the office or plant.
   d. A virus has infected your company's intranet and it will have to be shut down for twelve hours to debug it.
   e. The crane you are operating breaks down and you lose a half-day's work.
   f. The vendor shipped the wrong replacement part for your computer, and you cannot complete a job without buying a more expensive software package.
   g. An electrical storm knocked out your computer; you lost 1,000 mailing label addresses and will have to hire additional help to complete a mandatory mailing by the end of the week.
   h. A scammer has stolen some sensitive files (documents) about a new product your company hoped to launch next month.

7. Write a short internal proposal, similar to the one by Tina Escobar and Oliver Jabur in Figure 8.8 (pp. 262–265), recommending to a company or a college a specific change in procedure, equipment, training, safety, personnel, or policy. Make sure your collaborative writing team provides an appropriate audience (college administrator, department manager, or section chief) with specific evidence about the existence of the problem and your solution for it. Possible topics include

   a. providing more and safer parking
   b. forming a Usenet or listserv group
   c. purchasing new office or laboratory equipment or software
   d. hiring more faculty, student workers, or office help
   e. reorganizing or redesigning the school yearbook, company annual report, sales catalog, or website
   f. changing the decor/furniture in a student or company lounge
   g. increasing the number of online classes in your major
   h. adding more health-conscious offerings to the school or company cafeteria menu
   i. altering the programming on a campus radio station
   j. expanding distance learning offerings

8. Write a sales proposal, similar to the one in Figure 8.9 (pp. 269–271), on one of the following services or products you intend to sell or on a topic your instructor approves.

   a. providing exterminating service to a retail store or restaurant
   b. supplying a hospital with rental television sets for patients' rooms
   c. designing websites
   d. offering temporary office help or nursing care
   e. providing landscaping and lawn care service
   f. testing for noise, air, or water pollution in your community or neighborhood
   g. furnishing transportation for students, employees, or members of a community group, or individuals with disabilities
   h. supplying insurance coverage to a small firm (five to ten employees)
   i. making a work area safer
   j. selling a piece of equipment to a business
   k. offering discounted, or sponsored, memberships at a fitness center

CHAPTER 9

Writing Careful Long Reports

This chapter introduces you to long reports—and how and why they are written, organized, and documented. It is appropriate to discuss long reports in one of the last chapters of this book because they require you to use and combine many of the writing skills and research strategies you have already learned. In the world of global business, a long report can be the culmination of many weeks or months of hard work on an important company project.

To expand your understanding of writing long reports, take advantage of the Web Links, Additional Activities, and ACE Self-Tests at **college .cengage .com/pic/ kolinconcise2e.**

Characteristics of a Long Report

The following sections explain some of a long report's key elements. A model long report (Figure 9.1) appears at the end of the chapter (pp. 291–308).

Scope

A long report is a major study that provides an in-depth view of a key problem or idea. For example, a long report written for a course assignment may be eight to twenty pages long; a report for a business or industry may be that long or, more likely, much longer, depending on the scope of the subject. The implications of a long report are wide-ranging for a business or industry—relocating a facility, adding a new line of equipment, changing a computer programming operation.

The long report examines a problem in detail, while the short report covers just one part of the problem. Unlike a short report, a long report may discuss not just one or two current and routine events, but rather a continuing history of a problem or idea (and the background information necessary to understand it in perspective).

The titles of some typical long reports further suggest their extensive (and in some cases exhaustive) coverage:

- *A Master Plan for the Recreation Needs of Dover Plains, New York*
- *The Transportation Problems in Kingford, Oregon, and the Use of Monorails*
- *Promoting More Effective E-Commerce and E-Tailing at TechWorld*
- *The Use of Virtual Reality Attractions in Theme Parks in Jersey City, New Jersey*

> ■ *Public Policy Implications of Expanding Health Care Delivery Systems in Tate County*
>
> ■ *Internet Medicine in Providing Health Care in Rural Areas: Ways to Serve Southern Montana*

Research

A long, comprehensive report requires much more extensive research than a short report does. Information can be gathered over time from primary and secondary research—Internet searches, listservs, books, articles, laboratory experiments, on-site visits and tests, conferences with your boss and coworkers, interviews, and the writer's own observations. For a course report, you will have to do a great deal of research and possibly interviewing to track down the relevant background information and to discover what experts have said about the subject and what they propose should be suggested or even done. For a report for class you will be asked to identify a major problem or topic, while in business the topic and even your approach to it will more than likely be dictated to you by your boss and company policy.

Format

A long report is too detailed and complex to be adequately organized in a memo or letter format. The product of thorough research and analysis, the long report gives readers detailed discussions and interpretations of large quantities of data. To present the information in a logical and orderly fashion, the long report contains various parts, sections, headings, subheadings, documentation, and supplements (appendixes) that would never be included in a short report. The long report, as in Figure 9.1, gives readers a variety of visuals, including charts, a graph, and even a multicultural calendar.

Timetable

A long report is generally commissioned by a company or an agency to explore with extensive documentation a subject involving personnel, locations, costs, safety, or equipment. Many times a long report is required by law—for example, investigating the feasibility of a project that will affect the ecosystem. When you prepare a long report for a class project, select a topic that really interests you because you will spend a good portion of the term working on it. Here is a possible timetable for a long report.

Do research and conduct interviews	Outline or Draft	Conferences for Revisions and Further Research	Revise and Prepare Figures	Proof-read and Polish	Submit Long Report
4 weeks	3 weeks	1 week	3 weeks	1 week	Due Date

Audience

The audience for a long report is always individuals in the top levels of management—presidents, vice presidents, superintendents, directors—who make executive, financial, and organizational decisions. These individuals are responsible for long-range planning, seeing the big picture so to speak. A long report written about a campus issue or problem may at first be read by your instructor and then sent to an appropriate decision maker, such as a dean of students, a business manager, a director of athletics, or the head of campus security.

Collaborative Effort

The long report in the world of business may not be the work of one employee. Rather, it may be a collaborative effort—the product of a committee or a group. Your instructor may ask you to work in a group (or alone) in preparing your long report.

The group should estimate a realistic time necessary to complete the various stages of their work—when drafts are due or when editing must be concluded, for example. A project schedule based on that estimate should then be prepared. **But remember: Projects always take longer than initially planned.** Prepare for a possible delay at any one stage. The group may have to submit progress reports (see pp. 246–250) to its members as well as to management.

To be successful, a collaborative writing team should also observe the guidelines as well as the procedures for collaboration in the writing process discussed on pages 46–48.

The Process of Writing a Long Report

Because your work will be spread over many weeks, you need to see your report not as a series of static or isolated tasks but as an evolving project. Before you embark on that project, review the information on the writing process in Chapter 2. You may also want to review the flow chart in Figure 6.11 (p. 192), which illustrates the different stages in writing a research paper. The following guidelines will also help you plan and write a long report.

1. **Identify a significant topic.** You'll have to do some preliminary research—general reading, on-line searching, conferring with and interviewing experts—to get an overview of key ideas and individuals involved, and the implications for your company and/or community. Note the kinds of research Terri Smith Ruckel did for her long report (see Figure 9.1, pp. 291–308). As she did, expect to search a variety of print and Internet sources, to read and evaluate them, and to incorporate them into your work.

2. **Expect to confer regularly with your supervisor(s) and/or team members.** In these meetings, be prepared to ask pertinent and researched questions to pin down exactly what your boss wants and how your writing team can accomplish this goal.

3. **Revise your work often.** Be prepared to work on several outlines and drafts. Your revisions may sometimes be extensive, depending on what your boss, instructor,

or collaborative team recommends. You may have to consult new sources and arrive at a new interpretation of those sources.

4. Keep the order flexible at first. Even as you work on your drafts and revisions, remember that a long report is not written in the order in which the parts will finally be assembled. You cannot write in "final" order—abstract to recommendations. Instead, expect to write in "loose" order to reflect the process in which you gathered information and organized it for the final copy of the report. Usually, the body of the report is written first, the introduction later so that the authors can make sure they have not left anything out. The abstract, which appears very early in the report, is always written after all the facts have been recorded and recommendations are made or conclusions drawn.

5. Prepare both a day-to-day calendar and a checklist. Keep both posted where you do your work—above your desk or computer, or use your computer's built-in calendar program, if available—so that you can track your progress. The calendar should mark **milestones**—that is, dates by which each stage of your work must be completed. Match the dates on your calendar with the dates your instructor or employer may have given you to submit an outline, progress report(s), and the final copy. Your checklist should list the major parts of the long report. As you complete each section, check it off. Before assembling the final copy of your report, use the checklist to make sure you have not omitted something.

■ Parts of a Long Report

A long report may include some or all of the following twelve parts, which form three categories: *front matter* (letter of transmittal, title page, table of contents, list of illustrations, abstract), *report text* (introduction, body, conclusion, recommendations), and *back matter* (glossary, references cited, any appendixes).

Front Matter

As the name implies, the front matter of a long report consists of everything that precedes the actual text of the report. Such elements introduce, explain, and summarize to help the reader locate various parts of the report. Use lowercase roman numerals for front matter page numbers, not Arabic numbers.

Letter of Transmittal

This three- or four-paragraph (usually only one-page) letter states the purpose, scope, and major recommendation of the report. It highlights the main points of the report that your readers would be most interested in. If written to an instructor, the letter should additionally note that the report was done as a course assignment. (See Terri Smith Ruckel's letter on p. 291 for a sample letter of transmittal for a business report.)

Title Page

Find out what your boss or instructor prefers. Basically, your title page should contain the full title of your report and how you have restricted it in time, space, or method. Avoid titles that are vague, too short, or too long.

Vague Title:	A Report on the Internet: Some Findings
Too Short:	The Internet
Too Long:	A Report on the Internet: A Study of Dot-com Companies, Their History, Appeal, Scope, Liabilities, and Their Relationship to On-going Work Dealing with Consumer Preferences and Protection Within the Past Five Years in the Midwest

For a report for a class assignment, give your instructor's name and the specific course for which you prepared the report. For most other long reports, the title page needs

- the name(s) of the report writer(s)
- the date of the report
- the name of the firm or individual for whom the report was prepared

Table of Contents

The table of contents lists the major headings and subheadings of your report and tells readers on which pages they can be found. Essentially, a table of contents shows how you organized your report and emerges from many outlines and drafts. The items on those outlines frequently expand, shrink, combine, and can move around until you decide on the formal divisions and subdivisions of your report.

List of Illustrations

This list contains the titles for all of the visuals and indicates where they can be found in your long report.

Abstract

An abstract summarizes the report, including the main problem you investigated, the conclusions you reached, and any recommendations you may make. It is usually one paragraph long.

Abstracts may be placed at various points in long reports—on the title page, on a separate page, or as the first page of the report text but always in the front of the document.

The abstract may be the most important part of your report. Not every member of your audience will read your entire report, but almost everyone will read the abstract. For example, the president of the corporation or the director of an agency may use the abstract as the basis for approving the report and passing it on for distribution. Look at the abstract Terri Smith Ruckel prepared for her readers (p. 294).

Text of the Report

The text of a long report consists of an introduction, the body, conclusions, and sometimes recommendations.

Introduction

The introduction may constitute as much as 10 or 15 percent of your report, but it should not be any longer. The introduction is essential because it tells readers why your report was written and thus helps them to understand and interpret everything that follows. Avoid putting your findings, conclusions, or recommendations in your introduction.

Do not regard the introduction as one undivided block of information. It includes related parts, which should be labeled with subheadings. Here are the types of information to include in your introduction.

1. Background. To understand why your topic is significant, readers need to know about its history including information on who was originally involved, when, and where; how someone was affected by the issue; what opinions have been expressed on the issue; what the implications of your study are. See how the report on multinational employees (Figure 9.1, pp. 291–308) provides useful background information on when, where, how, and why these employees entered the U.S. work force and the effect of 9/11 on such workers and their employers.

2. Problem. Identify the problem or issue that led you to write the report. Because the problem you investigate will determine everything you write about in the report, you need to state it clearly and concisely. Here is a problem statement from a report on how construction designs have not taken into account the requirements of a growing number of seniors and disabled Americans.

> The construction industry has not satisfactorily met the needs for accessible workplaces and homes for all age and physical ability groups. The industry has relied on expensive and specialized plans to modify existing structures rather than creating universally designed spaces that are accessible to everyone.

3. Purpose statement. The purpose statement tells readers why you wrote the report and what you hope to accomplish or prove. It expresses the goal of all your research. Like the problem statement, the purpose statement does not have to be long or complex. A sentence or two will suffice. You might begin simply by saying, "The purpose of this report is . . ."

4. Scope. This section informs readers about the specific limits—number and type of issues, time, money, locations, personnel, and so forth—you have placed on your investigation. You inform readers about what they will find in your report, or what they won't, through your statement about the scope of your work. The long report in Figure 9.1 concentrates on adapting the U.S. workplace to meet the communication and cultural needs of a work force of multinational employees, not trends in the international market—two completely different topics.

Body

This section, also called the *discussion,* is the longest, possibly making up as much as 70 percent of your report. The body contains statistical information, details about the environment, and physical descriptions, as well as the various interpretations and comments of the authorities whose work you consulted as part of your research.

The body of your report should

- be carefully organized to reveal a coherent and well-defined plan
- separate material into meaningful parts to identify the major issues

- clearly relate the parts to each other
- use headings to help your reader identify major sections more quickly

Your organization should reflect the different headings (and even subheadings) included in your report; use them throughout to make it easy to follow. The headings, of course, will be included in the table of contents. (Note how Figure 9.1 is carefully organized into four sections.)

In addition to including headings at the beginning of each major section of the body, tell readers what they will find in that section and why. The report in Figure 9.1 effectively provides such internal summaries.

Conclusion(s)

The conclusion should tie everything together for readers by presenting the findings of your report. For a research report based on a study of sources located through various reference searches, the conclusion should summarize the main viewpoints of the authorities whose works you have cited. Perhaps your instructor will ask you to assess in your conclusion which resource materials were most thorough and helpful and why. For a marketing report done for a business, you must spell out the implications for your readers in terms of costs, personnel, products, location, and so forth.

Recommendations

The recommendation(s) section tells readers what should be done about the findings recorded in the conclusion. Your recommendation shows readers how you want them to solve the problem your report has focused on. Readers will expect you to advise them on a specific course of action—what equipment to purchase, when to expand a market, how to establish or revise a website or even an intranet, whom to hire for which positions or programs, and so on.

Back Matter

Included in the back matter of the report are all of the supporting data that, if included in the text of the report, would bog the reader down in details and cloud the main points the report makes.

Glossary

The glossary is an alphabetical list of the specialized vocabulary used in a long report and the definitions. A glossary might be unnecessary if your report does not use a highly technical vocabulary or if *all* members of your audience are familiar with the specialized terms you do use.

References Cited

Any sources cited in your report—websites, books, articles, television programs, interviews, reviews, audiovisuals—are usually listed in this section. Always ask your instructor or employer how he or she wants information to be documented.

Appendix

An appendix contains supporting materials for the report—tables and charts too long to include in the discussion, sample questionnaires, budgets and cost estimates, correspondence about the preparation of the report, case histories, transcripts of telephone conversations, copies of relevant letters and documents upon which the report is based, etc.

▌The Whys and Hows of Documentation

Documentation is an important component of any long report for at least three reasons.

1. It demonstrates that you have done your homework by consulting experts on the subject and relying on the most current and authoritative sources to build your case persuasively.
2. It gives proper credit to those sources. Citing works by name is not a simple act of courtesy; it is an ethical requirement and, because so much of the material is protected by copyright, a point of law.
3. It informs readers about specific books, articles, or websites you used so they can find your source and verify your facts or quotations.

The Ethics of Documentation: What Must Be Cited

To ensure that your paper or report avoids any type of plagiarism (see page 21) and maintains high ethical standards, follow these guidelines.

- If you use a source and take something from it, document it. Document any direct quotation(s), even a single phrase or keyword.
- Stay away from **patchworking**—using bits and pieces of information and passing them off as your own—which is also an act of plagiarism. Always put quotations marks around anything you take verbatim and document it.
- If any opinions, interpretations, and conclusions expressed verbally or in writing are not your own (e.g., you could not have reached them without the help of another source), document them.
- Even if you do not use an author's exact words but still get an idea, concept, or point of view from a source, document the work in your paper.
- Never alter any original material to have it suit your argument. Changing any information—names, dates, times, test results—is a serious offense.
- If you use statistical data you have not compiled yourself, document them.
- Do not use just one source or repeatedly cite the same source. You will not have sufficiently researched and documented your work for the reader.
- Always document any visuals—photographs, graphs, tables, charts, images downloaded from the Internet (and if you construct a visual based on someone else's data, you must acknowledge that source too).
- Never submit the same research paper you wrote for one course for another course without first obtaining permission from the second instructor.

- Do not delete an author's name when you are citing or forwarding an Internet document. You are obligated to give the Internet author full credit.

What Does Not Need to Be Cited

Be careful not to distract readers with unnecessary citations. They will demonstrate your lack of understanding of the documentation process and will undercut the professionalism of your report. You do not need to cite the following:

- Common-knowledge scientific facts and formulas, such as "The normal human body temperature is 98.6 degrees Fahrenheit," or "H_2O is the chemical formula of water."

- Well-known dates, such as the date of the first moon landing in 1969.

- Factual historical information, such as "George W. Bush was the 43rd president of the United States."

- Proverbs from folklore, such as "The hand is quicker than the eye."

- Well-known quotations, such as "We hold these truths to be self-evident . . . ," although it may be helpful to the reader if you mention the name of the person being quoted.

- The Bible, Koran, or other religious texts: Do not cite these on your Works Cited list, but do provide a parenthetical reference to the text and to the portion of the text quoted (for instance, *New Jerusalem Bible*, Exod. 2.3).

- Classic literary works: Do not cite these on your Works Cited list, but do provide a parenthetical reference to the original author and the name of the work. If the literary work is broken into parts, list the part quoted (for instance, Twain, *The Adventures of Huckleberry Finn*, Chapter 4, or Shakespeare, *The Merchant of Venice*, Act 2, Scene 3). Indicate, though, from which edition you took the quotation.

Parenthetical Documentation

Two frequently used systems of parenthetical documentation are found in the *MLA Handbook for Writers of Research Papers* and the *Publication Manual of the American Psychological Association.* The Modern Language Association (MLA) system is used primarily in the humanities and other related disciplines; the APA system is used in psychology, nursing and allied health disciplines, the social sciences, and in some business and technological fields. Both MLA and APA use parenthetical, or in-text, documentation. That is, the writer tells readers directly in the text of the paper, at the moment the acknowledgment is necessary, what reference is being cited.

MLA "Creating an effective website was among the top three priorities businesses have had over the last two years" (Morgan 203).

APA "Creating an effective website was among the top three priorities businesses have had over the last two years" (Morgan, 2008, p. 203).

The MLA citation "(Morgan 203)" or the APA "(Morgan, 2008, p. 203)" informs readers that the writer has borrowed information from a work by Morgan, specifically from page 203. APA also includes the year Morgan's work was published. Such a source (author's last name, year, and page number) obviously does not supply complete documentation. Instead, the parenthetical reference points readers to an alphabetical list of works that appears at the end of the report. The list, called "Works Cited" in MLA or "References" in APA, contains full bibliographic data—titles, dates, web addresses, publishers, page numbers, and so on—about each source cited in your report. Every work that appears in your report must be listed in your references section. To provide accurate parenthetical documentation for your readers, first carefully prepare your Works Cited or References list (see next subsection) so that you know which sources you are going to cite in the right form and at the right place in your text.

Keep your documentation brief and to the point so that you do not interrupt the reader's train of thought. In most cases, all you will need to include is the author's last name, date, and appropriate page number(s) in parentheses, usually at the end of sentences. When you mention the author's name in your sentence, though, MLA and APA both advise that you do not redundantly cite it again parenthetically; for example:

MLA   Moscovi claims that "Tourism has increased 21 percent this quarter" (76)

APA   Moscovi (2008) claims that "Tourism has increased by 21 percent this quarter" (p. 76)

For unsigned articles or radio and television programs, use a shortened title in place of an author's name parenthetically.

MLA   Shrewd bosses know that "chain-of-command meetings provide the opportunity to pass information up as well as down the administrative ladder" ("Working Smarter" 33).

APA   Shrewd bosses know that "chain-of-command meetings provide the opportunity to pass information up as well as down the administrative ladder" ("Working smarter, not harder," 2005, p. 33).

Similarly, if you list the title of a reference work in the text of your paper, do not repeat it in your documentation.

MLA   According to the *Encyclopedia Britannica,* Cecil B. DeMille's *King of Kings* was seen by nearly 800,000,000 individuals (3: 458).

APA   According to the *Encyclopedia Britannica* (2001), Cecil B. DeMille's *King of Kings* was seen by nearly 800,000,000 individuals (3, p. 458).

The first number in parentheses in both versions refers to the volume number of the *Encyclopedia Britannica;* the second is the page number in that volume.

Works Cited or References Pages

Examples of some of the entries you are likely to include in your References or Works Cited lists follow. Note the differences in the placement of information, punctuation, use of italics and quotation marks, and capitalization between MLA and APA. Regardless of which system you follow, arrange entries alphabetically and double-space within and between each entry.

Book by One Author

MLA   Spraggins, Marianne. *Getting Ahead: A Survival Guide for Black Women in Business*. Indianapolis: Wiley, 2009.

APA   Spraggins, M. (2009). *Getting ahead: A survival guide for black women in business*. Indianapolis: Wiley.

Book by Two Authors

MLA   Shaw, Timothy M., and Shaun Breslin. *China in the Global Political Economy*. New York: Palgrave Macmillan, 2007.

APA   Shaw, T., & Breslin, S. (2007). *China in the global political economy*. New York: Palgrave Macmillan.

Book by Three Authors

MLA   Brown, M. Katherine, Brenda Huettner, and Char James-Tanny. *Managing Virtual Teams: Getting the Most from Wikis, Blogs, and Other Collaborative Tools*. Plano: Wordware, 2007.

APA   Brown, M. K., Huettner, B., & James-Tanny, C. (2007). *Managing virtual teams: Getting the most from wikis, blogs, and other collaborative tools*. Plano, TX: Wordware.

Edited Collection of Essays

MLA   Martin, Jeanne B., and Robert C. Solomon, eds. *Honest Work: A Business Ethics Reader*. New York: Oxford, 2007.

APA   Martin, J., & Solomon, R. (Eds.). (2007). *Honest work: A business ethics reader*. New York: Oxford.

Works Included in a Collection of Essays

MLA   Sleet, David A., and Andrea Carlson Gielen. "Injury Prevention." *Health Promotion in Practice*. Ed. Sherri S. Gorfield. San Francisco: Jossey-Bass, 2006. 361–91.

APA   Sleet, D. A., & Gielen, A. C. (2006). Injury prevention. In S. S. Gorfield (Ed.), *Health promotion in practice* (pp. 361–391). San Francisco: Jossey-Bass.

Book by Corporate Author

MLA   Computer Literacy Foundation. *PCs in the Classroom*. 3rd ed. New York: Technology P, 2008.

APA   Computer Literacy Foundation. (2008). *PCs in the classroom* (3rd ed.). New York: Technology Press.

Article in a Professional Journal

MLA   Staples, D. Sandy, and Linda Zhao. "The Effects of Cultural Diversity in Virtual Teams Versus Face-to-Face Teams." *Group Decision and Negotiation* 15.4 (2006): 389– 406.

APA   Staples, D. S., & Zhao, L. (2006). The effects of cultural diversity in virtual teams versus face-to-face teams. *Group Decision and Negotiation, 15*,4, 389–406.

Article in a Print Magazine

MLA   Kadlec, Dan. "Rethinking Nonprofits." *Time* 5 Mar. 2007: 74.

APA   Kadlec, D. (2007, March 5). Rethinking nonprofits. *Time*, 74.

Article in a Professional Online Journal

MLA   Manojlovich, Marilyn. "Power and Empowerment in Nursing: Looking Backward to Inform the Future." *Online Journal of Issues in Nursing* 12.1 (2007). 31 Jan. 2007. <http://www.nursingworld.org/ojin/topic32/tpc32_1.htm>.

APA   Manojlovich, M. (2007). Power and empowerment in nursing: Looking backward to inform the future. *Online Journal of Issues in Nursing 12*(1). Retrieved January 31, 2007, from http://www.nursingworld.org/ojin/topic32/tpc32_1.htm

Article in a Newspaper

MLA   Hopkins, Jim. "How to Avoid Spam Avalanche." *USA Today* 21 Feb. 2007: B3.

APA   Hopkins, J. (2007, February 21). How to avoid spam avalanche. *USA Today*, p. B3.

Online Encyclopedia Article

MLA   David Ranson. "Inflation." *Concise Encyclopedia of Economics*. 2007. 15 Dec. 2007 <http://www.econlib.org/library/Enc/Inflation.html>.

APA   Ranson, D. (2007). Inflation. *Concise encyclopedia of economics*. Retrieved December 15, 2007, from http://www.econlib.org/library/Enc/Inflation.html

Online Anonymous Encyclopedia Article

MLA   "Iphone." *MSN Encarta Encyclopedia*. Microsoft 2007 <http://encarta.msn.com/encyclopedia_761552652_2/Apple_Inc.html#p21>

APA   Iphone. (2007). *MSN Encarta Encyclopedia*. Retrieved March 1, 2007, from http://encarta.msn.com/encyclopedia_761552652_2/Apple_Inc.html#p21

Unsigned Article in Print Magazine or Newspaper

MLA   "How Technology Could Create 'Eyes' for a Forklift." *Safety & Health* Jan. 2007: 18.

APA   How technology could create "eyes" for a forklift. (2007, January) *Safety & Health*, 18.

Website

MLA   National Council of La Raza. Home page. <http://www.nclr.org/>. 16 Aug. 2007.

APA   According to APA guidelines, websites do not need to be included on the Works Cited page, but URLs cited should appear in parentheses within your report.

Online Posting

MLA   Schwartz, Joseph, CEO, Sun Microsystems. "Sun Analyst Conference." Online posting. 8 Feb. 2007. 12 Feb. 2007 <http://blogs.sun.com/jonathan/>.

APA   Schwartz, J. (2007, February 8). Sun Analyst Conference. *Jonathan's Blog*, archived at http://blogs.sun.com/jonathan/entry/sun_analyst_conference.

Personal Interview

MLA   Alvarez, José. Professor of Physics, Northwest College. E-mail interview. 15 Oct. 2008.

APA   Interviews, conversations, and presentations are not included in APA reference lists, but you must still document them within your paper as follows: José Alvarez, Professor of Physics (personal communication, October 15, 2009).

E-Mail Correspondence

MLA   Teke, Cho-Martin. E-mail to the author. 3 July 2009.

APA   Treat as unpublished interview and cite parenthetically in the text.

Online Forum

MLA   Faoud, Jasine. "Procedures for New Nurses." Online posting. 26 June 2008. St. Adelbert's Hospital Home Page. 2 Sept. 2008 <http://www.saintadelberthome.com/>.

APA   Faoud, J. (2008, September 2). Procedures for new nurses. Message posted to http://www.saintadelberthome.com/

■ A Model Long Report

The following long report in Figure 9.1 (pages 291–308) was written by a senior training specialist, Terri Smith Ruckel, for her boss, the human resources director who commissioned it. Ruckel's main task was to demonstrate what U.S. businesses, and especially RPM, should do to meet the needs of multinational workers and thus promote diversity in the workplace. She gathered relevant data from both primary and secondary research, including journal articles, websites, interviews, government documents, and even in-house publications such as newsletters and company records, all accessed through the RPM intranet.

Figure 9.1 contains all the parts of a long report discussed in this chapter except a glossary and an appendix. Intended for a general reader interested in learning more about the problems multinational workers face, Ruckel's report does not contain the technical terms and data that might require a glossary and an appendix. In so doing, the report follows the American Psychological Association (APA) system of documentation. Yet even though the APA no longer recommends a table of contents or list of illustrations, we supply these pages as models for students who are asked to use them. Note how her cover letter introduces her report and its significance for RPM Technologies and her abstract succinctly identifies the main points of her report.

A long report.   **Figure 9.1**

RPM Technologies

4500 Florissant Drive   St. Louis, MO 63174

314.555.2121   www.rpmtech.com

May 3, 2007

Jesse Butler
Human Resources Director
RPM Technologies

Dear Director Butler:

With this letter I am enclosing my report on effective ways to recruit and retain a multinational work force for RPM Technologies, which you requested six weeks ago. My report argues for the necessity of adapting the RPM workplace to meet the needs of multinational employees, including promoting cultural sensitivity and making business communications more understandable.

Multinational workers undoubtedly are playing a major role in U.S. business and at RPM. With their technical skills and homeland contacts, they can help RPM Technologies successfully compete in the global marketplace.

Businesses like RPM need to recruit qualified multinational workers aggressively and then provide equal opportunities in the workplace. But we must also be sensitive to cultural diversity and communication demands. By including cross-cultural training—for native and non-native English-speaking employees alike—businesses like RPM promote cultural sensitivity. In-house language programs and plain-English or translated versions of key corporate documents can further improve the workplace environment for our multinational employees.

I hope you will find this report useful and relevant in your efforts to recruit additional multinational employees for RPM Technologies. If you have any questions or want to discuss any of our recommendations or research, please call me at extension 5406 or e-mail me at the address below. I look forward to your suggestions.

Sincerely yours,

Terri Smith Ruckel

Terri Smith Ruckel
Senior Training Specialist
truckel@rpmtech.com

Enclosure

Begins with major recommendation of report

Presents findings of report

Alerts reader to major issue

Offers to answer questions

Title page is carefully formatted and uses boldface for the title

Adapting the RPM Workplace for Multinational Employees in the New Millennium

Identifies writer and job title

Terri Smith Ruckel
Senior Training Specialist
RPM Technologies

Date submitted

May 3, 2007

RPM executive who assigned the report

Prepared for

Jesse Butler
Human Resources Director

2

Table of Contents

Major divisions of report in all capital letters

Subheadings indicated by different type-face, indentations, and italics

Page numbers included for major sections of report

3

List of Illustrations

Identifies each figure by title and page number

Abstract

Concise, informative abstract states purpose of report and why it is important

Uses helpful transitional words

This report investigates how U.S. businesses such as RPM must gain a competitive advantage in today's global marketplace by recruiting and retaining a multinational work force. This current wave of immigrants is in great demand for their technical skills and economic ties to their homeland. Yet many companies like ours still operate by policies designed for native speakers of English. Instead, we need to adapt our company policies and workplace environment to meet the cultural, religious, social, and communication needs of these multinational workers. To do this, we need to promote cultural sensitivity training, both for multinationals and employees who are native speakers of English. Additionally, like other U.S. firms, RPM should adapt vacation schedules and daycare facilities for its multicultural work force. Equally important, RPM should ensure, either through translations or plain-English versions, that all documents can be easily understood by multinational workers. We might also offer non-native speakers of English in-house language instruction while providing foreign language training for our employees who are native speakers of English.

4

Introduction

Background

The U.S. work force is undergoing a remarkable revolution. The Bureau of Labor Statistics predicts that by 2012 the labor force in the United States will comprise 162 million workers who must fill 167 million jobs (2006). According to the U.S. Chamber of Commerce, a shortage of skilled workers is sure to increase "even more heavily in the future when many of the baby boomers begin to retire" (2006).

The most dramatic effect of filling this labor shortage will be in hiring greater numbers of multinational employees, including those joining RPM. Currently, "one of every five computer specialists [and] one of every six persons in engineering or science occupations . . . is foreign born" (Kaushal & Fix, 2007). This new wave of immigrants will comprise 37 percent of the labor force by 2015 and continue to soar afterwards. By 2025 the number of international residents in the United States will rise from 26 million to 42 million, according to the U.S. Chamber of Commerce (2006). Even in light of immigration reforms mandated after 9/11, "over the next decade net immigration will average between 500,000 and 1.5 million people annually" (Congressional Budget Office, 2006). As Alexa Quincy aptly puts it, "the United States is becoming the most multiculturally diverse country in the global economy" (2007).

The following commissioned report explores the implications that this new multinational work force will have on RPM and what we must do to accommodate these workers.

Immigration: Then and Now

Although the United States has been seen as a nation of immigrants, the experiences of the current influx of new arrivals differ radically from those of their predecessors (Congressional Budget Office, 2006). The first great surge of immigration occurred in the late nineteenth and early twentieth centuries when nearly 9 million individuals entered the United States, mostly from the west and central European countries, as shown in Figure 1(a). Many of those citizens never went back to their homeland (Brown, 2007). But today's immigrants arrive from India, Pakistan, China, Mexico, Indonesia, the Philippines, and almost every other place around the globe, as Figure 1(b) reveals.

Actively maintaining ties with their native countries, these new immigrants travel back and forth so regularly they have become global citizens, exercising an enormous influence on a business like RPM. Demographers Crane and Boaz claim, "Immigration [will] give America an economic edge in the global economy . . . most notably in the Silicon Valley and other high-tech centers. They provide business contacts with other markets, enhancing [a company's] ability to trade and invest profitably abroad" (2005).

Gives convincing statistical evidence about the importance of topic

APA cites year of publication

Cites various sources to validate projections

Explains why the report was written

Contrasts immigration patterns and effects between an earlier period and now

Uses ellipses and brackets to show deletions/additions in quotes

5

*Numbers and
titles figures*

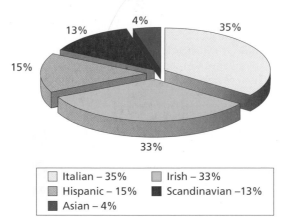

Figure 1(a)
**Major Ethnic Groups
Immigrating to New York City: 1900**

☐ Italian – 35%	☐ Irish – 33%
☐ Hispanic – 15%	■ Scandinavian –13%
■ Asian – 4%	

*Pie charts
dramatically
reveal differ-
ences in immi-
grant workers*

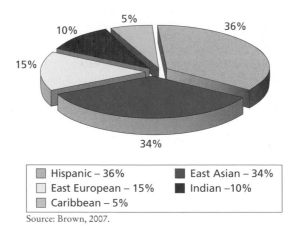

Figure 1(b)
**Major Ethnic Groups
Immigrating to New York City: 2006**

■ Hispanic – 36%	■ East Asian – 34%
☐ East European – 15%	■ Indian –10%
☐ Caribbean – 5%	

*Cites source for
visual*

Source: Brown, 2007.

*Describes new
workers and
documents
their signifi-
cance to
businesses*

High-Tech Immigrants

Undeniably, many immigrants today often possess advanced levels of technical expertise. A report by the Kaiser Foundation found that California's
Silicon Valley has significantly benefited from the immigrants who have arrived with much needed technical training. Asian, Indian, Pakistani, and
Middle Eastern scientists and engineers, who have relocated from a number
of countries, now hold more than 40 percent of the region's technical positions
("Immigrants attain," 2006). Figure 2 (p. 6) indicates the countries of origin
for Silicon Valley's immigrants and records each nationality's percentage.

6

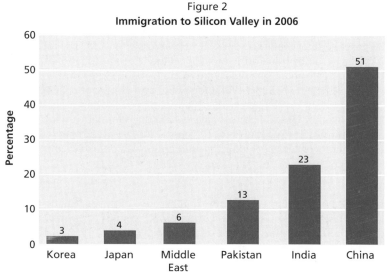

Figure 2
Immigration to Silicon Valley in 2006

Source: *Asian American Business Journal.*

   Despite a business downturn following the events of 9/11 and other
more recent terrorist activities in Madrid and London, information technology (IT) employers are enthusiastically searching for skilled international
workers. The annual immigration limit, set at 65,000 by the controversial
H-1B highly skilled worker visa, has been met on the very first day of a
company's fiscal year by many IT firms (Frauenheim, 2004). Murali
Krishna Devarakonda, president of the Immigrants Support Network in
Silicon Valley, states why: "We contribute significantly to the research and
development of a company's technology" (personal communication, April
3, 2007).

The Challenge to U.S. Businesses

As the U.S. population becomes more culturally diverse, RPM and many
other U.S. corporations will be challenged to organize and to conduct business in new ways. E-commerce companies like RPM are striving to adapt
their corporate policies and training programs to meet the communication
needs of this new sector of multilingual consumers and employees (Adamson, 2007). Annuncio Software in Los Altos may set the pace for the rest of
the U.S. corporate world. CEO Didier Moretti boasts about the high level of
cultural diversity among employees: "The last time we counted, we had 28
languages represented in our work force. I think having a diversity of backgrounds is a big help. Both small and large businesses . . . need to think
about hiring employees who can relate to customers in markets abroad"
(cited in Hamilton, 2007, p. 37). As Devarakonda concluded, "It is a win-win
situation. Multinational employees excel in the U.S. business culture, and
employers profit from their expertise and contacts."

7

Problem

Identifies a major problem and why it exists

Meeting the cultural and communication demands of this important labor force poses serious challenges for U.S. employers like RPM. The traditional workplace has to be transformed to respect the ways multinational employees communicate about business. Native English-speaking employees will also have to be better prepared to understand and appreciate their international co-workers. Unfortunately, many corporate policies and programs at RPM, and at other U.S. companies, have been created for native-born, English-speaking employees (Martin, 2007; Reynolds, 2007). Rather than rewarding multinational workers, such policies unintentionally punish them.

Two separate, corroborating sources

Purpose

Concisely states why the report was written

The purpose of this report is to show that because of the increasing numbers of multicultural employees in the workplace, RPM must adapt its business environment for this essential and diverse work force.

Scope

Informs reader that report will focus directly on RPM's needs

This report explores cultural diversity in the U.S. workplace in the new millennium and suggests ways for RPM to compete successfully in the global village by providing equal employment opportunities for multinational workers, fostering cross-cultural literacy, and improving training in intercultural communication.

Discussion

Providing Equal Workplace Opportunities for Multinational Employees

Discussion will be organized into three main headings, each with subheadings

Aggressive Recruitment of People from Diverse Cultures

A multilingual work force is essential in our culturally diverse global market. But many firms such as RPM must adapt or modify hiring policies and procedures to attract these multinational employees, beginning with rethinking our recruitment and retention policies. Routine visits to U.S. campuses by company recruiters or "specialized international recruiters" can help to identify and to recruit multinational job candidates (Hamilton, 2007, p. 35).

Emphasizes recruiting multinational workers and suggests how to do so

We should even consider visiting universities abroad with distinguished technical programs to attract talented multinational employees. These searches should be combined with our websites and executive blogs to emphasize RPM's commitment to globalization. Working more cooperatively with the Immigration and Naturalization Service will also help us to retain the most qualified, professional staff.

Identifies specific benefits for RPM

Capitalizing on a diverse work force, RPM can more effectively increase its multicultural customer base worldwide. Logically, customers buy from the people they can relate to culturally. RPM can take a lead from Union

8

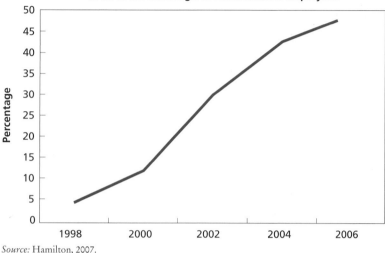

Figure 3
**Union Bank of California:
Growth in Percentage of Multinational Employees**

Source: Hamilton, 2007.

Tracks key information in a clear and concise graph

Bank of California, a business that effectively serves a diverse West Coast population, especially its Asian and Hispanic customers. The bank has a successful recruitment history of hiring employees with language skills in Japanese, Vietnamese, Korean, and Spanish. In fact, Union Bank is ranked fourth overall as an employer of minorities ("*Fortune* magazine," 2004). Figure 3 above charts the increase in the percentage of multinational employees hired by Union Bank.

Another successful business, Darden Restaurants, Inc., selected Richard Rivera, a Hispanic, to serve as president of Red Lobster, the nation's largest full-service seafood chain. Overseeing 680 restaurants nationwide, Rivera was the "most powerful minority in the restaurant industry" (R. Jackson, personal interview, January 15, 2007). Under Rivera's leadership, Red Lobster established an impressive history of hiring more international employees—totaling more than 35 percent of its work force by 2004—than it had in previous years. Rivera believes that management cannot properly respond to customers from different ethnic backgrounds if the majority of its employees is limited to native speakers of English. Closer to RPM in St. Louis, Whitney Abernathy—manager of Netshop, Inc.—found that contracts from Indonesia increased by 17 percent after she hired Jakarta native Safja Jacoef (personal communication, April 2, 2007).

Interview supports the need for recruitment

Interview not included in APA Reference list

Commitment to Ethnic Representation

Many companies have strong mission statements on diversity and multinational employees in the workplace. The most progressive of these companies—

9

Stresses other business precedents that proactively recruit these employees

Toyota, American Airlines, IBM, and Wal-Mart—promote multinationals as mentors and interpreters. Sodexho, which is committed to an all-inclusive workplace, has "five employee network groups, including 'SOL,' or the Sodexho Organization of Latinos" that has a "strong partnership with the National Council of Hispana Leadership" ("Top 50," 2006). Such a proactive program, which we would do well to incorporate at RPM, recognizes the leadership abilities of multinational employees. Moreover, "glass ceilings," which in the past have prevented women and ethnic employees from moving up the corporate ladder, are being shattered. The recruitment and promotion of non-native speakers of English is vital for corporate success. Tesfaye Aklilu, vice president at United Technologies, astutely observes:

Indent quotation of forty or more words

> In a global business environment, diversity is a business imperative. Diversity of cultures, ideas, perspectives and values is the norm of today's business environment. The exchange of ideas from different cultural perspectives gives businesses additional, valuable information. Every employee can see his/her position and value from a global point ("Diversity at UTC," 2007).

Cites programs offered by competitor

Exemplifying other innovative ways corporations recognize the importance of international employees, Ablex Resources, one of RPM's major competitors, recently created the position of Vice President of Diversity to establish a global work environment. And JP MorganChase "makes diversity an integral part of how it manages and does business" by holding its managers "responsible for administering a formal diversity plan . . . linking diversity to education, recruiting, and planning business growth" ("Top 50," 2006). For a second year in a row, Marriott International has been selected by *Diversity Inc.* magazine as one of the leading companies for "recruiting and retaining" Asian American and women managers (Marriott, 2007). American Airlines has also been honored for "using cultural differences to benefit employees and the company" (American Airlines, 2007).

Offers persuasive quotation found on company website

Second major section

Promoting and Incorporating Cultural Awareness Within the Company

Cross-Cultural Training

Cites business incentive to adapt as competitors did

References internal document from company intranet

Many of RPM's competitors are creating cultural awareness programs for international employees as well as native speakers. Employees find it easier to work with someone whose values and beliefs they understand, while employers benefit from effective on-the-job collaboration. Such a program could have prevented the problem last quarter when one of our non-native English-speaking employees was offended by a cultural misunderstanding ("RPM First Quarter," 2007). We may want to model our cultural sharing programs after those at American Express, which has a work force representing 40 nations, or those at Extel Communications with its large percentage of Hispanic and Vietnamese employees. United Parcel Service (UPS) profitably pairs a

10

native English-speaking employee with someone from another cultural group to improve on-the-job problem solving and communication. Jamie Allen, a UPS employee since 1995, found her work with Lekha Nfara-Kahn to be one of the most rewarding experiences of her job (cited in Adamson, 2007).

Although they need to encourage cultural sensitivity training, U.S. firms like RPM also need to be cautious about severing international workers' cultural ties—a delicate balance. When management actively promotes bonds among employees from similar cultures, workers are less fearful about losing their identity and becoming "token" employees. The law firm of Latour and Lleras recommends that new multinationals join a "buddy system" to assist their transition to American culture while retaining their cultural identity (U.S. Visa News, 2005). Globe Citizens Bank has long mobilized culturally similar groups by asking workers of identical cultural heritages to network with each other (Gordon & Rao, 2007, p. 36). Employees of Turkish backgrounds from Globe's main New York office go to lunch twice a month with Turkish-born employees from the Newark branches. Globe Citizens hosts these luncheons and in return receives a bimonthly evaluation of the bank's Turkish and Middle Eastern policies (Hamilton, 2007, p. 36).

Cultural education must go both ways, though. Both sides have things to learn about doing business in a global environment. The U.S. business culture has conventions, too, and few international employees would want to ignore them, but they need to know what those conventions are (Adamson, 2007). A frequent problem with U.S. corporations such as RPM is that we assume everyone knows how we do things and how we think—it never occurs to us to explain ourselves. Problems often stem from easy-to-correct misunderstandings about business etiquette. For example, native speakers of English are typically comfortable within a space of 1.5 to 2 feet for general personal interactions in business. But workers from Taiwan or Japan, who prefer a greater conversational distance, feel uncomfortable if their desks are less than a few feet away from another employee's workspace (Johnson, 2007, p. 44; "Taiwanese business culture," 2007).

Promotion of Cultural Sensitivity

Company efforts to respect different cultures might also include the recognition of an ethnic group's holidays or memorable historical events. RPM has just begun to do this by hosting cultural events, including Cinco de Mayo and Chinese New Year celebrations. Techsure, Inc., an Illinois software firm, allows Muslim employees to alter their work schedules during Ramadan, Islam's holy month of fasting (M. Saradayan, personal communication, February 28, 2007). Many companies have honored National Hispanic Heritage Month in September, which coincides with the independence celebrations of five Latin American countries (U.S. Department of Equal Employment Opportunity [USEEO], 2006). GRT Systems sends New Year's greetings at Waisak (the Buddhist Day of Enlightenment) to its Chinese employees and

Comment included in a source

Transition to new subdivision—networking of employees with similar cultural backgrounds

Relevant source on topic of immigration

Offers two examples RPM could follow

Another clear transitional sentence

Identifies key RPM problem

Gives cultural example

Gives precise ways RPM can incorporate cultural sensitivity into the workplace

11

Includes appropriate calendar of ethnic holidays

Gives page reference for visual

Identifies current RPM programs and how they could be easily modified to assist multinational workers

Cites company publication showing research within the organization

Third major section

Turns to written communication and multinational workers

Offers practical solution

to Indian workers at Rama Dipawli. Chemeka Taylor, GRT's operations manager, wisely points out: "We send native-born employees Christmas cards; why shouldn't we acknowledge our international work force, too?" (personal communication, April 10, 2007). Figure 4 (on page 12) provides a multicultural calendar that RPM needs to follow in developing cultural sensitivity policies.

Successful U.S. firms have been sensitive to the needs of their English-speaking employees for years. Flexible scheduling, telecommuting options, daycare, and preventative health programs have become part of corporate benefit plans to take care of employees. Many of these options and benefits have already been in place at RPM. But an international work force presents additional cultural opportunities for management to respond to with sensitivity. For example, company cafeterias might easily accommodate the dietary restrictions of vegetarian workers or those who abstain from certain foods, such as dairy products or meat. At Globtech, soybean and fish entrees are always available (Reynolds, 2007). Adding ethnic items at RPM would express our cultural awareness and respect for multinational employees.

Daycare raises critical issues for all working parents, native as well as non-native speakers of English. A discussion of daycare is also a key issue for RPM because we have offered such facilities at our high-tech park in San Luis Obispo, which has brought us much positive feedback over the past eight years ("RPM daycare facilities," 2007). By providing child care that reflects our workers' culturally diverse needs, RPM can give a multinational work force greater peace of mind and better enable them to do their jobs. DEJ Computers, for example, insists that at least two or three of its daycare workers must be fluent in Korean or Hindi (Parker, 2007). Another culturally sensitive employer, Mid-Plains Systems, assisted its Hispanic work force by hiring bilingual daycare workers and by serving foods the children customarily eat at home (Gordon & Rao, 2007, p. 39).

Making Business Communication More Understandable for Multinational Employees

Translation of Written Communications

All employees must be able to understand business communications affecting them. Among the essential documents causing trouble for multicultural readers are company handbooks, insurance and health care obligations, policy changes, and OSHA and EPA safety regulations (Hamilton, 2007). To ensure maximum understanding of these documents by a multinational work force, RPM will need to provide a translation or at least a plain-English version of them. To do this, RPM could solicit the help of employees who are fluent in the non-native English speakers' language as well as contract with professional translators to prepare appropriate documents.

12

Figure 4
A Multicultural Calendar

December 2008

	1	2	3	4	5	6
						Feast of St. Nicholas (some European countries)
7 Advent begins (Christian)	8 Bodhi Day (Rohatsu–Buddhist)	9 Eid al-Adha (Islam) Gita Jayanti (Hindu)	10	11	12 Datta Jayanti (Hindu); Feast Day—Our Lady of Guadalupe (Catholic)	13
14	15	16	17	18	19	20
21	22 Ekadashi (Hindu); Hanukkah begins (Jewish) (ends Dec. 29)	23	24	25 Christmas (Christian)	26 Kwanzaa begins (Interfaith) (ends Jan. 1)	27
28	29 Muharram (Islam)	30	31 New Year's Eve (Western)			

Some Information About December's Daily Observance
Bodhi Day: Buddhists celebrate Prince Gautama's vow to attain enlightenment.
Christmas: Christians celebrate the birth of Jesus.
Datta Jayanti: Celebration of the birth of Lord Datta.
Eid al-Adha: Feast of Sacrifice, commemorating the prophet Abraham's sacrifice of his son Ishmael to Allah.
Ekadashi: Twice monthly purifying fast.
Feast of Our Lady of Guadalupe: (Mexico) Catholic Christian holiday honoring the appearance of the Virgin Mary near Mexico City in 1531.
Feast of Saint Nicholas: Christmas celebration in Austria, Belgium, France, Germany, the Netherlands, Czech Republic, Poland, Russia, and Switzerland.
Gitta Jayanti: Celebration of the anniversary of the Bhagavad Gita, the Hindu scripture.
Hanukkah: Eight-day celebration commemorating the rededication of the Temple of Jerusalem.
Kwanzaa: African American and Pan-African celebration of family, community, and culture. Seven life virtues are presented.
Muharram: Islamic New Year.

Source: Johnson (2007): 45.

Major visual with a great deal of detail merits a full page to make it readable

Pays attention to major world holidays

Lists source

13

Figure 5
Safety Messages with Translations
for Multinational Employees

Danger: Radioactive

Peligro! Radiactivo (Spanish)

Dhamki! Jauhari
Tawaanaa'i! (Urdu)

Keikoku: Houshasei
busshitsu (Japanese)

Caution: Wet floor

Precaución! Mojado Suelo
(Spanish)

Khabar darrkarna!
Aabi farsh (Urdu)

Yuka ga nureteiru
(Japanese)

For your protection,
wear safety glasses

Pour votre protection,
verres de sûreté d'usure
(French)

Paheñna áasim chashmah
(Urdu)

Anzen megane wo chakuyoo
shitekudasai (Japanese)

Source: Parker, 2007.

Cites examples of safety visuals with translations into multinational workers' native languages

Workplace signs in particular, especially safety messages, must consider the language needs of international workers. In the best interest of corporate safety, RPM needs to have these signs translated into the languages represented by multinationals in the workplace and/or to post signs that use global symbols, such as those shown in Figure 5 above. Unquestionably, we need to avoid signs that workers might find hard or even impossible to decipher. For example, a large **P** for "parking" or a ⍰ for "questions are answered here" or an **H** for "hospital" might be unfamiliar to non-native speakers of English (Parker, 2007).

Gives examples, and what to avoid and why

Exchange of Language Learning

Outlines specific benefits for RPM

Providing English language instruction for employees also has obvious advantages for the RPM workplace in the new millennium. Should RPM elect not to provide in-house language training, the company should consider reimbursing employees for necessary courses, books, tapes, and software. Moreover, we might make available an audio library of specific vocabulary and phrases commonly used on the job as well as multilingual dictionaries with relevant technical terms. Hiromi Naguchi, an e-marketing analyst at PowerUsers Networking, praised the English language training she received on her job. Although she had ten years of English in the Osaka (Japan)

Points to apt example

14

schools, Naguchi developed her listening and speaking skills on the job, benefiting her employers, co-workers, and customers. As a result of her diligence, Naguchi was recognized as her company's "Most Valuable Employee for 2006" for her interpersonal skills (H. Naguchi, personal communication, April 3, 2007).

But language training has to be reciprocal—for native as well as nonnative speakers—if communication is to succeed in the global marketplace. Sadly, most Americans do not know a foreign language and "fewer than 1 in 10 students at U.S. colleges major in [a] foreign language" (Cutshall, 2005, p. 20). Many international workers are bi- or even trilingual. In India and South Africa, for example, the average employee speaks three or more national languages every day to conduct business. Stockard International boasts that its employees know more than 70 languages, everything from Albanian to Yoruba.

Argues that reader must consider both sides

Provides compelling statistics to prove point

A recent international survey of executive recruiters (2007) showed that being bilingual is critical to success in business ("Developing foreign language"). According to this survey, 85 percent of European recruiters, 88 percent of Asian recruiters, and 95 percent of recruiters in Latin America stressed the importance of speaking at least two languages in the new global village. As Jennifer Torres, CEO of Pacifica Corp., emphasized, "Although English remains the dominant language of international business, multilingual executives clearly have a competitive advantage. This advantage will only increase with the continued globalization of commerce" (cited in "Developing foreign language," 2007). RPM executives might want to contract with one of the companies specializing in language instruction for business people found at *http://www.selfgrowth.com/foreignlanguage.html.*

Cites business survey to confirm the necessity of change at RPM

Identifies quote within a source

Conclusions

To compete in the global marketplace, RPM, like other U.S. companies, must emphasize cultural diversity in its corporate mission and in the workplace. Through its policies and programs, RPM must aggressively recruit and retain an increasing number of technologically educated and experienced multinational workers who are in great demand today and will be even more so in the next ten to twenty years. These workers can help us increase our international customer base and advance the state of our technology. But we need to ensure that the workplace is sensitive to their cultural, religious, and communication needs. Providing equal opportunities, diversity training and networking, and easy-to-understand business documents will keep RPM globally competitive in recruiting and retaining these essential employees. We can expand only by making a commitment to learn about and respect other cultures and their ways of communicating and doing business.

Following the body, conclusion concisely summarizes the highlights of the report without repeating the documentation

15

Recommendations

Provides specific, relevant recommendations, based on conclusions, to solve the problem at RPM

Numbered list format is easy to read for busy executives

By implementing the following recommendations, RPM Technologies can succeed in hiring and promoting the IT multinational professionals our company requires.

1. Recruit multinational workers more effectively through our website, international hiring specialists, and visits to campuses here and abroad, and work more closely with the Immigration and Naturalization Service (INS) to retain multinationals.

2. Create cultural sensitivity and networking groups comprising multinational and native English-speaking employees. Develop educational materials for employees who are native speakers of English about the cultures of their multinational co-workers.

3. Establish a mentoring program to identify and foster leadership abilities in multinational employees.

4. Supply relevant translations and plain-English versions of company handbooks, manuals, new regulations, insurance policies, and safety codes.

5. Reassess and adapt our daycare facilities to meet the needs of children of multinational employees.

16

REFERENCES

Adamson, R. (2007, February). Challenges ahead for American business. *National Economics Review, 11*, 45–46.

American Airlines. (2007). International diversity awards. Retrieved March 5, 2007. http:www.aa.com/content/amrcorp/pressReleases/2006_06_0608_diverse.jhtml

Brown, P. (2007, December 15). History of U.S. immigration. Retrieved Feb. 2, 2007, from http://immigration.ucn.edu

Chan, M. (2007). Incident report (RPM Technologies report number 14P-C). RPM Technologies.

Congressional Budget Office. (2006, June). Projections of net immigration to the U.S. Retrieved February 20, 2007, from http://www.cbo.gov/ftpdocs/72xx/doc7249/0606/Immigration.pdf

Crane, E. H., & Boaz, D. (Eds.). (2005). *Cato Handbook on Policy* (6th ed.). Washington, DC: Cato.

Cutshall, S. (2005, January). Why we need "the year of languages." *Educational Leadership, 62*(4), 20–23.

Developing foreign language skills is good business. (2007, February 3). *Business World.* Retrieved April 11, 2007, from http://www.businessworld.ca/article.cfm/newsID/7109.cfm

Diversity at UTC. (2007). *United Technologies Corporation.* Retrieved March 5, 2007, from http://www.utc.com/careers/diversity/index4.htm

Fortune magazine ranks Union Bank one of the best companies for minorities. (2004, June 23). *Union Bank of California.* Retrieved December 12, 2006, from http://www.uboc.com/about/main/0,3250,2485_11256_502261585,00.html

Frauenheim, E. (2004, October 1). H-1B visa limit for 2005 already reached. *C-Net News.* Retrieved March 20, 2007, from http://news.com.com/H-1B+visa+limit+for+2005+already+reached/2100-1022_3-5392917.html?tag=nl

Hamilton, B. E. (2007, April 15). Diversity is the answer for today's workforce. *The New Business Journal, 10*(38), 35–37.

Immigrants attain the American dream in Silicon Valley. (2006, December 19). *The Taipei Times*, p. 12.

Johnson, V. M. (2007). Growing multinational diversity in business sparks changes. *Business Across the Nation, 23*(7), 43–48.

Marriott International. (2007, January 2). Awards and recognition—diversity. Retrieved January 30, 2007, from http://marriott.com/news/corporate-Profile/awardsDiversity.mi

Labels list of sources actually cited

All entries listed alphabetically by author's last name, or (if no author) by first word of title

In-house publication

Date of publication included for every entry after author's name or title if author's name not given

Only first word and proper nouns capitalized in title

Pages provided for print sources

APA requires access date as well as posting date in Web entries

16

Martin, P. (2007). Migration news. Retrieved January 20, 2007, from http://www.migration.ucdavis.edu

Parker, M. (2007, May). Immigration facts. Retrieved March 21, 2007, from http://immigration.org

Reynolds, P. (2007, March 30). Serving up culture. Retrieved April 21, 2007, from http://www.culture.org

RPM daycare facilities rated high. (2007, January 15). *RPM News*, 31–32. (Also available on RPM intranet)

RPM first quarter activity report. (2007). *RPM Internal Reports*. Retrieved June 4, 2007, from RPM intranet.

Taiwanese business culture. (2007). *Executive Planet*. Retrieved February 28, 2007, from http://www.executiveplanet.com/business-culture-in/132438266669.html

Top fifty companies for Hispanics in 2006. (2006, September). *Hispanic Business*. Retrieved February 16, 2007, from http://www.hispanicbusiness.com/news/newsbyid

U.S. Bureau of Labor Statistics. (2007). Tomorrow's jobs. Retrieved March 7, 2007, from http://www.bls.gov/oco/oco2003.htm

U.S. Chamber of Commerce. (2006, June). Chamber, labor leaders renew call for immigration reform. *USChamber.com*. Retrieved February 15, 2007, from http://www.uschamber.com/issues/index/immigration/uscc_migrationpolicy_statement/htm

U.S. Department of Equal Employment Opportunity. (2006). *Hispanic employment program report for 2005* (EEO Publication No. B56.2245). Washington, DC: U.S. Government Printing Office.

U.S. Visa News. (2005). Latour and Lleras immigration attorneys. Retrieved March 8, 2007, from http://www.usvisanews.com/relbuddy.shtml

Second and subsequent lines indented 5 spaces

Works without author alphabetized by first word of title

Full Web addresses are given for verification and to make source easy to find

✔ Revision Checklist

- [ ] Concentrated on a major problem—one with significant implications for my major, neighborhood, city, or employer.
- [ ] Identified, justified, and described the significance of the major problem as opposed to focusing on a minor or side issue.
- [ ] Did sufficient research—in the library, on the Internet, through interviewing, from personal observation and/or testing—to convince readers that I am knowledgeable about this problem, its scope and effects, and likely solution.
- [ ] Made sure I understand what my employer/instructor wants me to include.
- [ ] Followed company's/instructor's guidelines for style, visuals, and format.
- [ ] Consistently used one method of documentation that my company/instructor wanted me to follow, e.g., APA, MLA, or another model.
- [ ] Adhered to company's/instructor's schedule for completing various parts of the long report.
- [ ] Divided and labeled the parts of long report to make it easy for readers to follow and to show a careful plan of organization.
- [ ] Wrote a one-page letter of transmittal or cover letter informing readers why the report was written and describing its scope and findings.
- [ ] Supplied an abstract that leaves no doubt in readers' minds about what report deals with and why.
- [ ] Designed attractive title page that contains all the basic information—title, date, for whom the report is written, my name—readers require.
- [ ] Gave readers all the necessary introductory information about background, problem, purpose of report, and scope.
- [ ] Included in body of report the results of all my research—the facts, statistics, and descriptions—that my readers need in order to know that I have done my homework on the topic well.
- [ ] Wrapped up report in succinct conclusion. Told readers what the findings of my research are and accurately interpreted all data.
- [ ] Supplied a recommendations section (if required) that tells readers concretely how they can respond to the problem using the data. Recommendations make sense—are realistic and practical and related directly to the research and topic. Ensured that recommendations are persuasive.
- [ ] Included in the final copy of report all the parts listed in table of contents.

Exercises

Additional Activities related to writing long reports are located at **college.cengage .com/pic/ kolinconcise2e.**

1. What kinds of research did the employee do to write the long report in Figure 9.1?

2. Study Figure 9.1 and answer the following questions based on it.
   a. How has the writer successfully limited the scope of the report?
   b. Where and how does the writer use visuals especially well?
   c. Where and how has the writer adapted her technical information for her audience (a general reader)?
   d. What visual devices does the writer use to separate parts of the report and divisions within each part?
   e. How does the writer introduce, summarize, and draw conclusions from the expert opinions she cites in order to substantiate the main points?
   f. What are the ways in which the writer documents information she has gathered?
   g. What functions does the conclusion serve for readers? Cite specific examples from the report.
   h. How and why do the recommendations follow from the material presented in the report?

3. As a team or alone, come to class prepared to discuss at least two major problems that would be suitable topics for a long report. Consider an important community problem—traffic, crime, air and water pollution, housing, transportation—or a problem at your college. Then write a letter to a consulting firm, or other appropriate agency or business, requesting a study of the problem and a report.

4. Write a report outline for one of the problems you decided on in Exercise 3. Use major headings and include the kinds of information discussed in the Front Matter section of this chapter (pp. 280–281).

5. Have your instructor look at and approve the outline you prepared for Exercise 4. Then write a long report based on the outline, either individually or as part of a collaborative writing team.

Making Successful Presentations at Work

Almost every job requires employees to have and to use carefully developed speaking skills. In fact, to get hired, you have to be a persuasive speaker at your job interview. And to advance up the corporate ladder, you will have to continue to be a confident, well-prepared, and persuasive speaker. The goal of this chapter is to help you be successful in your presentations at work and before clients.

To expand your understanding of making presentations, take advantage of the Web Links, Additional Activities, and ACE Self-Tests at **college .cengage .com/pic/ kolinconcise2e.**

Types of Presentations

On the job you will have numerous speaking responsibilities that will vary in the amount of preparation they require, the time they last, and the audience and the occasion for which they are intended. Here are some frequent types of presentations you can expect to make as part of your job:

- sales appeals to prospective customers
- evaluations of products or policies
- progress reports to your boss and clients
- reports to superiors about your job accomplishments
- findings of your long report
- justification of your position or even your department
- appeals and/or explanations before elected officials
- presentations at professional conferences
- explanation of a procedure, decision, or plan before a community/civic group (Chamber of Commerce or local PTA)

Whatever type of presentation you are asked to deliver, this chapter gives you practical advice on how to become a better, more assured speaker in both informal briefings and formal presentations.

Informal Briefings

If you have ever given a book report or explained laboratory results in front of a class, you have given an informal briefing. Such semiformal reports are a routine part of many jobs. Here are some of the typical informal briefings you may be asked to deliver at work:

- a status report on your current project
- an update or end-of-shift report, like those nurses and police officers give
- an explanation of a policy to co-workers
- a report on a conference you attended
- a demonstration of new equipment or software
- a follow-up session on equipment or training procedures
- a summary of a meeting you attended

Such presentations are usually short (one to seven minutes, perhaps), and you won't always be given advance notice. When the boss tells you to "say a few words about the new website" (or the new programming procedure), you will not be expected to give a lengthy formal speech.

Guidelines for Preparing Informal Briefings

Follow these guidelines when you have to make an informal briefing.

- Make your comments brief and to the point.
- Keyboard a few bulleted items you plan to cover.
- Highlight key phrases and terms you need to stress.
- Include in your notes only the major points you want to mention.
- Arrange your points in chronological order or from cause to effect.

Figure 10.1 is an informal outline with key facts used by an employee who is introducing Diana J. Rizzo, a visiting speaker, to a monthly meeting of safety directors.

■ Formal Presentations

Whereas an informal briefing is likely to be short, generally conversational, and intended for a limited number of people, a formal presentation is much longer, far less conversational, and may be intended for a wider audience. It also involves much more preparation.

Avoid speaking "off the cuff." The worst way to make a presentation is to speak without any preparation whatsoever. You only fool yourself if you think you will have all the necessary details and explanations in the back of your head. Once you start talking, generally everything does not fall into place smoothly. Without preparation, you are likely to confuse important points or forget them entirely. Mark Twain's advice is apt here: "It takes three weeks to prepare a good impromptu speech."

Expect to spend several days preparing your presentation. You cannot just dash it off. You will need time to

- research the subject
- interview key resource individuals
- prepare, time, and sequence visuals

- Diana J. Rizzo, Chief Engineer of the Rhode Island State Highway Department for twelve years

- Experience as both a civil engineer and safety expert

- Consultant to Secretary Habib, Department of Transportation

- Member of the National Safety Council and author of "Field Test Procedures in Highway Safety Construction"

- Instrumental in revising state Safety Commission's website

- Received "Award for Excellence" from the Northeastern Association of Traffic Engineers in May 2009

- coordinate your talk with presentations by co-workers or your boss
- rehearse your presentation

Many of us are uncomfortable in front of an audience because we feel frightened or embarrassed. Much of that anxiety can be eased if you know what to expect. The two areas you should investigate thoroughly before you begin to prepare your presentation are (1) who will be in your audience and (2) why they are there.

Analyzing Your Audience

The more you learn about your audience, the better prepared you will be to give them what they need. Just as you do for your written work, for your oral presentation you will have to do some research about the audience, emphasizing the "you attitude" and establishing your own credibility.

Consider Your Audience as a Group of Listeners, Not Readers

While audience analysis pertains both to readers of your work and to listeners of your presentation, there are several fundamental differences between these two groups. Unlike a reader of your report, the audience for your presentation

- is a captive group of listeners
- may have only one chance to get your message
- has less time to digest what you say
- has a shorter attention span
- can't always go back to review what you said or jump ahead to get a preview
- can be more easily distracted—by interruptions, chairs being moved, people coughing, and so on
- cannot absorb as many of the technical details you would include in a written report

Take all of these differences into account as you plan your presentation and assess who constitutes your audience.

Who Is Your Audience?

Here are five key questions to ask when analyzing your audience and their needs.

1. **How much do they know about your topic?**

   - consumers with little or no technical knowledge
   - technical professionals who understand terms, jargon, and background
   - business managers looking only for the bottom line

2. **What unites them as a group?**

   - members of the same profession/organization
   - customers using the same products
   - employees of the company you work for

3. **What is their interest level in your topic?**

   - highly motivated
   - neutral—waiting to be informed, entertained, or persuaded
   - uninterested in your topic—only there because attendance is mandatory
   - uncooperative and antagonistic, likely to challenge you

4. **What do you want them to do after hearing your presentation?**

   - buy a product or service
   - change a schedule
   - learn more about your topic
   - sign a petition

5. **What questions are they likely to raise?**

   - about money
   - about personnel/schedule
   - about locations

Special Considerations for a Multinational Audience

Given the international make-up of audiences at many business presentations, you may have to address a group of listeners whose native language is not English or even make a presentation before individuals in a country other than your own. Consider your audience's particular cultural taboos and protocols. Do they accept your looking at them directly, or do they frown on eye contact?

As you prepare a talk before a multinational audience, keep the following points in mind:

1. Brush up on your audience's culture, especially accepted ways they communicate with each other (see pp. 117–125).
2. Find out what constitutes an appropriate length for a talk before your audience.
3. Be especially careful about introducing humor—avoid anything that is based on nationality, dialect, religion, or race.
4. Think twice about injecting anything autobiographical into your speech. Some cultures regard such intimacy as an invasion of privacy.
5. Steer clear of politics; you risk losing your audience's confidence.
6. Choose visuals with universally understood icons.

Chapter 10 Additional Activities, located at **college.cengage .com/pic/ kolinconcise2e**, include a global-focused activity titled "Making Presentations for International Audiences."

▌The Parts of Formal Presentations

As you read this section, refer to Marilyn Claire Ford's PowerPoint presentation in Figure 10.2. Note how effectively she used PowerPoint (see pp. 316–318) to convince a potential client, GTP Systems, to purchase a service contract provided by World Tech, her employer. Her talk consists of seven slides that contain relevant images and concise text.

The Introduction

The most important part of a presentation is your introduction, which should capture the audience's attention by answering these questions: (1) Who are you? (2) What are your qualifications? (3) What specific topic are you speaking about? and (4) How is the topic relevant to us?

Your first and most immediate goal is to establish rapport with your audience, win their confidence, and elicit their cooperation. Since your audience is probably at their most attentive during the first few minutes of your presentation, they will pay close attention to everything about you and what you say. Seize the moment and build momentum.

An effective introduction should be proportional to the length of your presentation. A ten-minute speech requires no more than a sixty-second introduction; a twenty-minute speech needs no more than a two- or three-minute introduction. Note how Marilyn Claire Ford in slides 1 and 2 introduces herself, her company, and its benefits for GTP.

Figure 10.2   A sample PowerPoint presentation.

Introduces speaker, provides reason for presentation

Type size and font are clear

Switching to Videoconferencing: A Wise Choice for GTP

Marilyn Claire Ford
World Tech
Desktop Videoconferencing
November 15, 2008

Chooses short title for each slide

Succinctly lists benefits in short, easy-to-read bulleted points

What We Offer

- User-friendly desktop videoconferencing
- Cutting-edge communication technology
- Flexible low-cost networking
- Single network for data, voice, and video

First of four slides that make up the body of the presentation

Develops first key sales feature

Visual does not mask text

Easy to Use

With a simple telephone call you can
- Arrange a meeting with colleagues around the globe
- Use your computer to access World Tech's conferencing system
- See, hear, and talk with all participants

Continued

Cost Effective

- Dramatically reduces costs for travel
- Upgrades current computer system for less than the use of an existing computer
- Cuts data-processing expenses by 60%

Second point gives only essential facts

Does not overwhelm listeners with numbers

Appropriate icon emphasizing "cutting" costs

Improves Staff Efficiency

- Brings people together at the right time
- Enhances communication when employees see and hear each other
- Whiteboard technology aids collaboration

Third sales feature appropriately mentions specific technology

Leaves generous margins

Icon shows bringing staff together

Advantages in the Global Marketplace

- Global sales increase
- Clear recording of all meetings
- Safety risks lessened from decreased employee travel

Addresses security as last, most emphatic point in list of advantages

Figure 10.2   (Continued)

Conclusion

Issues a call to action; makes contact easy through website, e-mail, and telephone

> # Conclusion
>
> - Recap of technology benefits
> - Thank you for listening
> - Please sign up this week--it is easy
> - Just log on to www.WorldTech.com
> - Any questions?
>
> **Mcf@WorldTech.com**
> **1-800-271-5555**

How to Begin

You can begin by introducing yourself, emphasizing your professional qualifications and interests. (A self-introduction is unnecessary if someone else has introduced you or if you know everyone in the room.)

Give Listeners a Road Map

Give listeners a "road map" at the beginning of your presentation so that they will know where you are, where you are going, and what they have to look forward to or to recall. Indicate what your topic is and how you have organized what you have to say about it.

> My presentation today on Digital Business software will last about 20 minutes and is divided into three parts. First, I will outline briefly recent software changes. Second, I want to give you a detailed review of how those changes directly affect our company. Third, I will show how our company can profitably implement those changes. At the end of my presentation, there will be time for your questions and comments.

The most informative presentations are the easiest to follow. Restrict your topic to ensure that you will be able to organize it carefully and sensibly—for example, a tasty diet under 1,000 calories a day or a course in learning Java or another software package.

Capture the Audience's Attention

Use any of the following strategies to get your audience to "bite the hook."

- Ask a question. "Did you know that every fifteen minutes a foreign-owned business opens in China?"
- Start with a quotation. Winston Churchill said, "We get things to make a living but we give things to have a life." (Consult *Bartlett's Familiar Quotations* online at *http://www.bartleby.com* for a list of handy quotations.)

- Give an interesting statistic. "In 2008, 2 million heart attack victims will live to tell about it." (Go to the *World Almanac and Book of Facts* to find something relevant to your presentation topic.)
- Relate an anecdote or story. Be sure it is relevant and in good taste; make your audience feel at ease and friendly toward you by establishing a bond with them.

Again, be careful about using humor in a business talk. It could backfire—the audience may not get the point or may even be offended by it.

The Body

The body is the longest part of your presentation, just as it is in a long report. Make it persuasive and relevant to your audience by (1) explaining a process, (2) describing a condition, (3) solving a problem, (4) arguing a case, or (5) doing all of these. See how the body of Marilyn Claire Ford's presentation in Figure 10.2 is organized around the customer benefits of GTP's switching to desktop videoconferencing. In slides 3–6 she outlines how easy, economical, and efficient such technology is to use in a global marketplace.

To get the right perspective, recall your own experiences as a member of an audience. How often did you feel bored or angry because a speaker tried to overload you with details or could not stick to the point?

Ways to Organize the Body

Here are a few helpful ways you can present and organize information in the body of your presentation. In writing a report, design your document to help readers visually, supplying headings, underscoring, bullets, necessary white space, and headers and footers. In a presentation, switch from those purely visual devices to aural ones, such as the following:

1. Give signals (directions) to show where you are going or where you have been. Enumerate your points: *first, second, third.* Emphasize cause-and-effect relationships with *subsequently, therefore, furthermore.* When you tell a story, follow a chronological sequence and fill your speech with signposts: *before, following, next, then.*

2. Comment on your own material. Tell the audience if some point is especially significant, memorable, or relevant. "This next fact is the most important thing I'll say today."

3. Provide internal summaries. Spending a few seconds to recap what you have just covered will reassure your audience and you as well.

> I have already discussed the difficulties in establishing a menu repertory, or the list of items that the food service manager wants to appear on the menu. Now we will turn to ways of determining which items should appear on a menu and why.

4. Anticipate any objections or qualifications your audience is likely to have. Address the issues with relevant facts about costs, personnel, and/or equipment in your presentation.

The Conclusion

Plan your conclusion as carefully as you do your introduction. Stopping with a screeching halt is as bad as trailing off in a fading monotone. An effective conclusion should leave the audience feeling that you and they have come full circle and accomplished what you promised. Notice how Marilyn Claire Ford ends her presentation by persuading GTP Systems that they will gain a competitive edge, the point with which she began (see pp. 316–317).

What to Put in a Conclusion

A conclusion should contain something memorable. Never introduce a new subject or simply repeat your introduction. A conclusion can contain the following:

- a fresh restatement of your three or four main points
- a call to action, just as in a sales letter—to buy, to note, to agree, to volunteer
- a final emphasis on a key statistic (for example, "The installation of the stainless steel heating tanks has, as we have seen, saved our firm 32 percent in utility costs, since we no longer have to run the heating system all day.")

End your presentation, as in slide 7, with a concise summary of the main points and urge listeners to purchase your product or service.

Mean It When You Say, "Finally"

When you tell your audience you are concluding, make sure you mean it. Saying, "In conclusion," and then talking for another ten minutes frustrates listeners and makes them less receptive to your message.

▌ Presentation Software

As Marilyn Claire Ford's presentation in Figure 10.2 demonstrates, business talks frequently use PowerPoint, Corel Presentation, or other presentation software. Using these graphics packages is a crucial skill your employer will expect you to have. These software packages enable you to create an electronic slide show with concise text and carefully chosen visuals that can be created on any computer that has Windows 2000, XP, or Vista. They allow you to plan, write, and create your slides with visuals all at once.

To get the benefits of this technology, map your presentation before you actually create your slides. First, identify the large topics you want to cover, as Marilyn Claire Ford did in presenting information to save GTP time and money, increase staff efficiency, and help the company compete in a global economy. Then organize your topics logically and persuasively. Include only those supporting details that relate directly to your topic and to your audience's needs. These key points should help you determine the number of slides and visuals you use. But don't overwhelm audiences with too much information. Note that Marilyn Claire Ford needs only seven slides in Figure 10.2. Also choose your visuals carefully. Resist the temptation to dazzle your audience with electronic special effects. Your goal is not to create a glitzy show but to represent your company professionally.

Guidelines on Using Presentation Software Effectively

Here are some tips to ensure that the design, organization, and delivery of your PowerPoint presentation go smoothly. Refer to Figure 10.2 again as you study them.

Readability

- Make sure each slide is easy to read—clear and uncluttered
- Use a type size that is crisp and easy to see, even from a distance. For a small presentation on your laptop, use 30-point type or larger. Increase your type size for headings, as in Figure 10.2.
- Keep your type style and size consistent. Don't switch from one font to another.
- Avoid ornate or script type, and do not put everything in boldface or in all capital letters. Note how only the headings in Figure 10.2 are boldfaced.

Text

- Keep text short and simple. Use easy-to-recall names, words, and phrases.
- Use bulleted lists instead of unbroken paragraphs. But put no more than five bulleted lines on a slide and limit each line to seven or eight words. Don't squeeze words on a line. Include no more than 40 to 45 words per slide.
- Double-space between bulleted items and leave generous margins on all sides.
- Title each slide—using a question, a statement, or key name or phrase.

Sequencing Slides

- Keep your slides in the correct order in which you need to show them.
- Limit your presentation to twelve slides, maximum.
- Retain the same transition (cover left or straight right) from slide to slide to avoid visual confusion.
- Spend about two to three minutes per slide, but don't read each slide verbatim. Summarize main ideas or concisely expand them while looking at your audience, not the slide.
- Time your slides so your audience can read them. Never continue to show a slide after you have moved on to a new topic.
- If you invite audience participation and interaction, build in extra time between your slides.

Background/Color

- Find a pleasant contrasting background to make your text easy to read. Avoid extremely light or dark backgrounds that make it hard to read (see pp. 177–178).
- Use the same background for each slide.
- Avoid shadowing your text for "decorative" visual effect.
- Use color sparingly. Don't turn each slide into a sizzling neon sign.

Graphics

- Keep graphics clear, simple, and positioned appropriately on the slide.
- Be sure visuals do not cover or shadow text.
- Show only those visuals that support your main points. Not every slide requires a visual. Slides 1 and 7 in Figure 10.2, for example, do not use visuals.
- Include only one graphic per slide; otherwise, your text will be more difficult to read.
- Import visuals from identified sources rather than designing new ones.
- Avoid complicated tables, elaborate flow charts, or busy diagrams.
- Use clip art sparingly. Remember: Less is more. Clip art must be functional, not distracting. Each icon in Figure 10.2 is functional.
- Incorporate animation, sound effects, or movie/video clips only when they are persuasive, relevant, and undeniably professional.
- Don't bother with borders; they do not make a slide clearer.

Quality Check

- Be sure your spelling, grammar, names, dates, costs, and sources are correct.
- Double-check all math, tables, equations, and percentages.
- Don't show a slide if you aren't going to discuss it in your presentation.

Noncomputerized Presentations

You can expect to incorporate visuals in a variety of other media besides Power-Point presentations. There will be situations where you may have to use a conventional chalkboard, a flip chart (where large pieces of white paper are anchored on an easel and flipped over like pages in a notebook or tablet), or an overhead projector where you make your transparencies on sheets of clear plastic. You will almost surely be asked to prepare handouts that include visuals accompanying a presentation and distribute them either before or during your talk.

Regardless of the medium you use, make sure your visuals are

- visible
- easy to understand
- self-explanatory
- relevant
- accurate

Getting the Most from Your Noncomputerized Visuals

The following practical suggestions will help you get the most from your visuals when time and space may prohibit using computer setups.

1. Do not set up your visuals before you begin speaking. The audience will be wondering how you are going to use them and so will not give you their full attention. When you are finished with a visual, put it away so that your audience will not be distracted by it or tempted to study it instead of listening to you.

2. Firmly anchor any maps or illustrations.   Having a map roll up or a picture fall off an easel during a presentation is embarrassing.

3. Never obstruct the audience's view by standing in front of your visuals. Use a pointer or a laser pointer (which can project a bright red spot up to 150 feet) to direct the audience's attention to your visual.

4. Avoid crowding too many images onto one visual that you transfer to a projector or onto one pasteboard.   Use different screens or pasteboards instead.

5. Do not put a lot of writing on a visual.   Elaborate labels or wordy descriptions defeat your reason for using the visual. If any writing must appear on one of your visuals, enlarge it so your audience can read it quickly and easily.

6. Be especially cautious with a slide projector.   Check beforehand to make sure all your slides are in the correct order and are right side up. Most important, make sure the projector is in good working order. Practice changing from one transparency to another. And test your tape recorder if you are using one as part of your presentation.

▌Rehearsing Your Presentation

Don't skip rehearsing your talk thinking it will save you time. It will actually help you become more familiar with your topic and overall message, building your confidence. Here are some strategies to use as you rehearse your speech.

- Know your topic and the various parts of your presentation.
- Speak in front of a full-length mirror for at least one rehearsal to see how an audience might view you.
- Talk into a tape recorder to determine whether you sound friendly or frantic, poised or pressured. You can also catch and correct yourself if you are speaking too quickly or too slowly. A rate of about 120 to 140 words a minute is easy for an audience to follow.
- Time yourself so that you will not exceed your allotted limit or fall far short of your audience's expectations.
- Practice with the PowerPoint or other presentation software, visuals, equipment, or projector that you intend to use for valuable hands-on experience.
- Monitor the type of gestures (neither too many nor too few) you can use for clarity and emphasis in your talk.
- Videotape your final rehearsal and play it to a colleague or instructor for feedback.

▌Delivering Your Presentation

A poor delivery can ruin a good presentation. You will be evaluated by your style of presentation just as you are in your written work. When you speak before an audience, you will be evaluated on the image you project: how you look, how you talk, and how you move (your body language). Do you mumble into your notes, never looking at the audience? Do you clutch the lectern as if to keep it in place?

First impressions are crucial. Research shows that people decide what they think of you in the first two or three minutes of your presentation. The way you dress is important but so is your body language. In fact, 75 percent of your audience's impressions are influenced by your body language. The nonverbal signals you send affect how your audience will regard your leadership abilities, your sales performance, even your sincerity. Pay attention to gestures, movement of your hands, how you stand, and so on. No matter how many hours you have worked to get your message ready, if your nonverbal presentation is misleading or inappropriate, the impact of what you say will be lost.

The following suggestions on how to deliver a presentation will help you to be a well-prepared, poised speaker.

Settling Your Nerves Before You Speak

Being nervous before your presentation is normal—a faster heartbeat, sweaty palms, shaking. But don't let your nerves stand in your way of delivering a highly successful talk. Here are some ways you can calm yourself before you deliver your presentation and get some healthy doses of confidence.

- Give yourself plenty of time to get there. The more you have to rush, the more anxious you will be.
- Don't bring anything with you that is likely to spill, such as coffee, a soft drink, or bottled water.
- Avoid caffeine for a few hours before your talk if caffeine makes you jittery.
- Take some deep breaths and then hold your breath while you count to ten. Then exhale. This will slow your heart rate and lower your blood pressure.
- Remind yourself that you have spent hours preparing. Your hard work will pull you through.
- Try to chat with one or two members of the audience ahead of time and relax. See your audience as friends—people who can help your career.

Making Your Presentation

Everyone is nervous before a talk. Accept that fact and even allow a few seconds for "panic time." Then put your nervous energy to work for you. Chances are, your audience will have no idea how anxious you are; they cannot see the butterflies in your stomach. Again, see your audience as friends, not enemies. Remember to do the following:

1. **Establish eye contact with your listeners.**   Look at as many people in your audience as possible to establish a relationship with them. Never bury your head in notes or keep your eyes fixed on a computer screen or keyboard. You will only signal your lack of interest in the audience or your fear of public speaking. In a small PowerPoint presentation, try to establish rapport with each person in the room.

2. **Adjust to audience feedback.**   Watch your listeners' reactions and respond appropriately—nodding to agree, pausing a moment, paraphrasing to clarify a confusing point. Know your material well so that if someone asks you a question or

wants you to return to a point, you are not fumbling through your notes or trying feverishly to locate the right screen.

3. Use a friendly, confident tone.   Speak in a natural, pleasant voice, but avoid verbal tics ("you know," "I mean," "like he goes") and fillers ("um," "ah," "er") repeated several times each minute. Use pauses instead.

4. Vary the rate of your delivery.   Use your natural speaking voice. Vary your rate and inflection to help you emphasize key points and make transitions. Talk slowly enough for your audience to understand you, yet quickly enough so that you don't sound as if you are belaboring or emphasizing each word.

5. Adjust your volume appropriately.   Talking in a monotone, never raising or lowering your voice, will lull your audience to sleep or at least inattention! Talk loudly enough for everyone to hear, but be careful if you are using a microphone. Your voice will be amplified, so if you speak loudly, you will boom rather than project. Every word with a *b*, *p*, or *d* will sound like an explosive in your listener's ear. Watch out for the other extreme—speaking so softly that only the first two rows can hear you.

6. Watch your posture.   Don't shift from one foot to another. But do not slouch or look wooden either. If you stand motionless, looking as if rigor mortis has set in, your speech will be judged cold and lifeless, no matter how lively your words are. Be natural yet dynamic; smile, nod your head. Refer to an object on the screen by touching it, or use a pointer to emphasize something on an overhead projector.

7. Use appropriate body language.   Be natural and consistent. Do not startle an audience by suddenly pounding on the lectern or desk for emphasis. Avoid gestures that will distract or alienate your audience. For example, don't fold your arms as you talk, a gesture that signals you are unreceptive (closed) to your audience's reactions. Also, avoid the nervous habits that can divert the audience's attention: clicking a ballpoint pen, scratching your head, rubbing your nose, twirling your hair. Nor do you have to remain still or step with robotlike movements. The remote control for a PowerPoint presentation allows you to casually walk around the room as you click and change screens.

8. Dress professionally.   Do not wear flashy clothes or clanking jewelry (such as necklace or bracelet charms) that call attention to themselves. Wear clothes that are the business norm—know your company's dress code. Women should wear a businesslike dress or suit; men should wear a dark business suit or sports coat, white or blue shirt, and a tasteful tie.

Evaluating Presentations

A large portion of this chapter has given you information on how to construct and deliver a formal presentation. As a way of reviewing that advice, study Figure 10.3—an evaluation form similar to those used by instructors in communication classes. Note that the form gives equal emphasis to the speaker's performance or delivery and to the organization, content, and sequence of the presentation.

Figure 10.3   An evaluation form for an oral presentation.

Name of speaker _____   Date of presentation _____

Title of presentation _____   Length of presentation _____

PART I: THE SPEAKER (circle the appropriate number)

1. Appearance:	1 sloppy	2	3	4	5 well groomed
2. Eye contact:	1 poor	2	3	4	5 effective
3. Voice:	1 monotonous	2	3	4	5 varied
4. Posture:	1 poor	2	3	4	5 natural
5. Gestures:	1 disturbing	2	3	4	5 appropriate
6. Self-confidence:	1 nervous	2	3	4	5 poised

PART II: THE PRESENTATION (circle the appropriate number: 1 = poor; 5 = superior)

1. Speaker's knowledge of the subject—carefully researched; factual errors; missing details:

   1   2   3   4   5

2. Relevance of the topic for audience—suitable for this group:

   1   2   3   4   5

3. The speaker's language—too technical; filled with clichés or slang expressions; or crisp and descriptive:

   1   2   3   4   5

4. The speaker's slides and/or other visuals—easy to read, large, relevant, carefully timed and ordered:

   1   2   3   4   5

5. Presentation easy to follow—speaker gave the audience signs where he or she had been and where he or she was going:

   1   2   3   4   5

6. Presentation fit time limit; included appropriate number of slides:

   1   2   3   4   5

7. Speaker's conclusion—clearly identified major points:

   1   2   3   4   5

Revision Checklist

☐ Anticipated audience's background, interest, or even potential resistance, and questions about message of both informal and formal presentations.

☐ Prepared introduction to provide "road map" of presentation and to arouse audience interest.

☐ Started with interesting and relevant statistics, a question, an anecdote, or similar "hook" to capture audience attention.

☐ Limited body of presentation to main points.

☐ Sequenced main points logically and made connections among them.

☐ Used supporting examples and illustrations appropriate to audience.

☐ Made sure conclusion contains summary of main points of my presentation and/or specific call for action.

☐ Designed visuals that are clear, easy to read, and relevant for audience.

☐ Prepared relevant number of slides with key bulleted items and, where appropriate, icons.

☐ Experimented successfully with presentation software and applications before including it in a presentation.

☐ Rehearsed presentation thoroughly to become familiar with its organization and visuals.

☐ Monitored volume, tone, and rate to vary delivery and emphasize major points.

☐ Rehearsed gestures to make them relevant and nonintrusive.

☐ Timed presentation, complete with visuals, to run close to allotted time.

Exercises

1. Prepare a three- to five-minute presentation explaining how a piece of equipment that you use on your job works. If the equipment is small enough, bring it with you to class. If it is too large, prepare an appropriate visual or two for use in your talk.

2. You have just been asked to talk about the students at your school or the employees where you work. Narrow the topic and submit an outline to your instructor, showing how you have limited the topic and gathered and organized evidence. Use two or three appropriate visuals (tables, photographs, maps, charts, icons, or even videos) to give a PowerPoint presentation. Follow the format of the presentation in Figure 10.2.

3. Prepare a ten-minute presentation on a controversial topic that you would present before a civic group—the PTA, the local chapter of an organization, a post of the Veterans of Foreign Wars, a synagogue, a mosque, or a church club.

Additional Activities related to making presentations are located at college.cengage.com/pic/kolinconcise2e.

4. Using the information contained in the long report on multinational workers in Chapter 9 (pp. 291–308), prepare a short presentation (five to seven minutes) for your class.

5. Using the evaluation form in Figure 10.3, evaluate a speaker—a speech class student, a local politician, or a co-worker delivering a report at work. Specify the time, place, and occasion of the speech. Pay special attention to any visuals the speaker uses.

6. Deliver a formal presentation (fifteen to twenty minutes) on the various websites for individuals in your chosen career field. Restrict your topic and divide it into four key issues or parts, as in Figure 10.2. Use at least three visuals with your talk. Submit a written outline to your instructor.

A Writer's Brief Guide to Paragraphs, Sentences, and Words

To write successfully, you must know how to create effective paragraphs, write and punctuate clear sentences, and use words correctly. This guide succinctly explains some of the basic elements of clear and accurate writing.

■ Paragraphs

Writing a Well-Developed Paragraph

A paragraph is the basic building block for any piece of writing. It is (1) a group of related sentences (2) arranged in a logical order (3) supplying readers with detailed, appropriate information (4) on a single important topic.

A paragraph expresses one central idea, with each sentence contributing to the overall meaning of that idea. The paragraph does that by means of a *topic sentence*, which states the central idea, and *supporting information*, which explains the topic sentence.

Supplying a Topic Sentence

The topic sentence is the most important sentence in your paragraph. Carefully worded and restricted, it helps you to generate and control your information. An effective topic sentence also helps readers grasp your main idea quickly. As you draft your paragraphs, pay close attention to the following three guidelines.

   1. **Make sure you provide a topic sentence.**   In their rush to supply readers with facts, some writers forget or neglect to include a topic sentence. The following paragraph, with no topic sentence, shows how fragmented such writing can be.

   No topic sentence: Sensors found on each machine detect wind speed and direction and other important details such as ice loading and potential metal fatigue. The information is fed into a small computer (microprocessor) in the nacelle (or engine housing). The microprocessor automatically keeps the blades turned into the wind, starts and stops the machine, and changes the pitch of the tips of the blades to increase power under varying wind conditions. Should any part of the wind turbine suffer damage or malfunction, the microprocessor will immediately shut the machine down.

Only when a suitable topic sentence is added—"The MOD-2 wind turbine is programmed to run by the latest computer technology"—can readers understand what the technical details have in common.

2. Put your topic sentence first.   Place your topic sentence at the beginning—not the middle or end—of your paragraph because the first sentence occupies an emphatic position. Burying the key idea in the middle or near the end of the paragraph makes it harder for readers to comprehend your purpose or act on your information.

3. Be sure your topic sentence is focused.   If restricted, a topic sentence discusses only one central idea. A broad or unrestricted topic sentence leads to a shaky, incomplete paragraph for two reasons.

- The paragraph will not contain enough information to support the topic sentence.
- A broad topic sentence will not summarize or forecast specific information in the paragraph.

The following example of a carefully constructed paragraph contains a clear topic sentence in an appropriate position (highlighted in color) and adequate supporting details.

> Fat is an important part of everyone's diet. It is nutritionally present in the basic food groups we eat—meat and poultry, dairy products, and oils—to aid growth or development. The fats and fatty acids present in those foods ensure proper metabolism, thus helping to turn what we eat into the energy we need. Those same fats and fatty acids also act as carriers for important vitamins like A, D, E, and K. Another important role of fat is that it keeps us from feeling hungry by delaying digestion. Fat also enhances the flavor of the food we eat, making it more enjoyable.

Knowing the Three Characteristics of an Effective Paragraph

Effective paragraphs have **unity**, **coherence**, and **completeness**.

Unity

A unified paragraph sticks to one topic without wandering. Every sentence, every detail, **supports**, **explains**, or **proves** the central idea. A unified paragraph includes only relevant information and excludes unnecessary or irrelevant comments.

Coherence

In a coherent paragraph all sentences flow smoothly and logically to and from each other like the links of a chain. Use the following three techniques to achieve coherence.

1. Use transitional words and phrases.   Some useful connective, transitional words, along with the relationships they express, are listed in Table A.1.

> **Paragraph with connective words:** Advertising a product on the radio has many advantages over using television. *For one thing*, radio rates are much cheaper. *For example*, a one-time 60-second spot on television can cost $750. *For that money*, advertisers can purchase nine 30-second spots on the radio. *Equally attractive* are the low production costs for radio advertising. *In contrast*, television advertising often includes extra costs for models and voice-overs. *Another* advantage radio offers advertisers is immediate scheduling. *Often* the ad appears during the same week a contract is

TABLE A.1   Transitional, or Connective, Words and Phrases

Addition	again	besides	moreover
	additionally	first, second, third	next
	along with	furthermore	together with
	also	in addition	too
	and	many	what's more
	as well as		
Cause/effect	accordingly	consequently	on account of
	and so	due to	since
	as a result	hence	therefore
	because of	if	thus
Comparison/ contrast	but	in contrast	on the other hand
	conversely	in the same way	similarly
	equally	likewise	still
	however	on the contrary	yet
Conclusion	all in all	in brief	on the whole
	altogether	in conclusion	to conclude
	as we saw	in short	to put into perspective
	at last	in summary	to summarize
	finally	lastly	to wrap up
Condition	although	granted that	provided that
	depending	if	to be sure
	even though	of course	unless
Emphasis	above all	for emphasis	of course
	after all	indeed	surely
	again	in fact	to repeat
	as a matter of fact	in other words	to stress
	as I said	obviously	unquestionably
Illustration	for example	in other words	that is,
	for instance	in particular	to demonstrate
	in effect	specifically	to illustrate
Place	across from	below	over
	adjacent to	beyond	there
	alongside of	here	under
	at this point	in front of	where
	behind	next to	wherever
Time	afterward	formerly	previously
	at length	hereafter	soon
	at the same time	later	simultaneously
	at times	meanwhile	subsequently
	beforehand	next	then
	currently	now	until
	during	once	when
	earlier	presently	while

signed. *On the other hand*, television stations are *frequently* booked up months in advance, so it may be a long time *before* an ad appears. *Furthermore*, radio gives advertisers a greater opportunity to reach potential buyers. *After all*, radio follows listeners everywhere—in their homes, at work, and in their cars. *Although* television is very popular, it cannot do that.

2. Use pronouns and demonstrative adjectives.   Words like *he, she, him, her, they*, and so on contribute to paragraph coherence and increase the flow of sentences.

> Paragraph with pronouns:
> Traffic studies are an important tool for store owners looking for a new location. These studies are relatively inexpensive and highly accurate. They can tell owners how much traffic passes by a particular location at a particular time and why. Moreover, they can help owners to determine what particular characteristics the individuals have in common. Because of their helpfulness, these studies can save owners time and money and possibly prevent financial ruin.

3. Use parallel (coordinated) grammatical structures.   Parallelism means using the same *kind of* word, phrase, clause, or sentence to express related concepts.

> Orientation sessions accomplish four useful goals for trainees. First, they introduce trainees to key personnel in accounting, data processing, maintenance, and security. Second, they give trainees experience logging into the database system, selecting appropriate menus, editing core documents, and getting off the system. Third, they explain to trainees the company policies affecting the way supplies are ordered, used, and stored. Fourth, they help trainees understand their responsibilities in such crucial areas as computer security and use.

Parallelism is at work on a number of levels in the paragraph above, among them

- The four sentences about the four goals start in the same way grammatically ("... they introduce/give/explain/help ...") to help readers categorize the information.
- Within individual sentences, the repetition of *present participles* (log*ging*, se*lecting*, edit*ing*, get*ting*) and of *past participles* (order*ed*, us*ed*, stor*ed*) helps the writer to coordinate information.
- Transitional words—*first, second, third, fourth*—provide a clear-cut sequence.

Completeness

A complete paragraph provides readers with sufficient information to **clarify, analyze, support, defend,** or **prove** the central idea expressed in the topic sentence. The reader feels satisfied that the writer has given necessary details.

> Skimpy paragraph:
> Farmers can turn their crops and farm wastes into useful, cost-effective fuels. Much grown on the farm can be converted to energy. This energy can have many uses and save farmers a lot of money in operating expenses.

> Fully developed paragraph:
> Farm crops and wastes can be turned into fuels to save farmers on their operating costs. Alcohol can be distilled from grain, sugar beets, potatoes, even blighted crops. Converted to gasohol (90 percent gasoline,

10 percent alcohol), this fuel can run such farm equipment as irrigation pumps, feed grinders, and tractors. Similarly, through a biomass digestion system, farmers can produce methane from animal or crop wastes as a natural gas for heating and cooking. Finally, cellulose pellets, derived from plant materials, become solid fuel that can save farmers money in heating barns.

Sentences

Constructing and Punctuating Sentences

The way you construct and punctuate your sentences can determine whether you succeed or fail in the world of work. Your sentences reveal a lot about you. They tell readers how clearly or how poorly you can convey a message. And any message is only as effective and as thoughtful as the sentences of which it is made.

What Makes a Sentence

A sentence is a complete thought, expressed by a subject and a verb that can make sense standing alone.

> subject   verb
> Websites   sell   products.

The Difference Between Phrases and Clauses

The first step toward success in writing sentences is learning to recognize the difference between phrases and clauses. A **phrase** is a group of words that does not contain a subject and a verb; phrases cannot make sense standing alone. Phrases cannot be sentences.

in the park	No subject:	Who is in the park?
	No verb:	What was done in the park?
for every patient in intensive care	No subject:	Who did something for every patient?
	No verb:	What was done for the patients?

A **clause** does contain a subject and a verb, but *not every clause is a sentence.* Only **independent** (or **main**) **clauses** can stand alone as sentences. Here is an example of an independent clause that is a complete sentence.

> subject verb   object
> The president closed the college.

A **dependent** (or **subordinate**) **clause** also contains a subject and a verb, but it does not make complete sense and cannot stand alone. Why? A dependent clause contains a subordinating conjunction—*after, although, as, because, before, even though, if, since, unless, when, where, whereas, while*—at the beginning of the clause. Such conjunctions subordinate the clause in which they appear and make the clause dependent for meaning and completion on an independent clause.

After
Before
Because } the president closed the college
Even though
Unless

"After the president closed the college" is not a complete thought but a dependent clause that leaves us in suspense. It needs to be completed with an independent clause telling us what happened "after."

dependent clause	independent clause		
	subject	verb	phrase
After the president closed the college,	we	played	in the snow.

Avoiding Sentence Fragments

An incomplete sentence is called a **fragment**. Fragments can be phrases or dependent clauses. They either lack a verb or a subject or have broken away from an independent clause. A fragment is isolated: It needs an overhaul to supply missing parts to turn it into an independent clause or to glue it back to an independent clause to have it make sense.

To avoid writing fragments, follow these rules. **Note that incorrect examples are preceded by a minus sign, correct revisions by a plus sign**.

 1. **Do not use a subordinate clause as a sentence.**   Even though it contains a subject and a verb, a subordinate clause standing alone is still a fragment. To avoid this kind of sentence fragment, simply join the two clauses (the independent clause and the dependent clause containing a subordinating conjunction) with a comma—*not* a period or semicolon.

 – Unless we agreed to the plan. (What would happen?)
 – Unless we agreed to the plan; the project manager would discontinue the operation. (A semicolon cannot set off the subordinate clause.)
 + Unless we agreed to the plan, the project manager would discontinue the operation.
 – Because safety precautions were taken. (What happened?)
 + Because safety precautions were taken, ten construction workers escaped injury.

Sometimes subordinate clauses appear at the end of a sentence. They may be introduced by a subordinate conjunction, an adverb, or a relative pronoun (*that, which, who*). Do not separate these clauses from the preceding independent clause with a period, thus turning them into fragments.

 – An all-volunteer fire department posed some problems. Especially for residents in the western part of town.
 + An all-volunteer fire department posed some problems, especially for residents in the western part of town. (The word *especially* qualifies posed, referred to in the independent clause.)

 2. **Every sentence must have a subject telling the reader who does the action.**

 – Being extra careful not to spill the solution. (Who?)
 + The technician was being extra careful not to spill the solution.

3. Every sentence must have a complete verb. Watch especially for verbs ending in *-ing*. They need another verb (some form of *to be*) to make them complete.

> – The machine running in the computer department. (Did what?)

You can change that fragment into a sentence by supplying the correct form of the verb.

> + The machine *is running* in the computer department.
> + The machine *runs* in the computer department.

Or you can revise the entire sentence, adding a new thought.

> + The machine running in the computer department processes all new accounts.

4. Do not detach prepositional phrases (beginning with *at, by, for, from, in, to, with*, and so forth) **from independent clauses.** Such phrases are not complete thoughts and cannot stand alone. Correct the error by leaving the phrases attached to the sentence to which they belong.

> – By three o'clock the next day. (What was to happen?)
> + The supervisor wanted our reports by three o'clock the next day.

Avoiding Comma Splices

Fragments occur when you use only bits and pieces of complete sentences. Another common error that some writers commit involves just the reverse kind of action. They weakly and wrongly join two complete sentences (independent clauses) with a comma as if those two sentences were really only one sentence. Such an error is called a **comma splice**. Here is an example.

> – Gasoline prices have risen by 10 percent in the last mont<u>h, </u>we will drive the car less often.

Two independent clauses (complete sentences) exist:

> + Gasoline prices have risen by 10 percent in the last month.
> + We will drive the car less often.

A comma alone lacks the power to separate independent clauses.

As the preceding example shows, many pronouns—*I, he, she, it, we, they*—are used as the subjects of independent clauses. A comma splice will result if you place a comma instead of a semicolon between two independent clauses where the second clause opens with a pronoun.

> – Maria approved the pla<u>n, </u>she liked its cost-effective approach.
> + Maria approved the pla<u>n; </u>she liked its cost-effective approach.

However, relative pronouns (*who, whom, which, that*) are preceded by a comma, not a period or a semicolon, when they introduce subordinate clauses.

> – She approved the plan. Which had the cost-effective approach.
> + She approved the plan, which had the cost-effective approach.

Four Ways to Correct Comma Splices

1. Remove the comma separating two independent clauses and replace it with a period.   Then capitalize the first letter of the first word of the new sentence.

+ Gasoline prices have risen by 10 percent in the last month. We will drive the car less often.

2. Insert a coordinating conjunction (*and, but, or, nor, so, for, yet*) **after the comma.**   Together, the conjunction and the comma properly separate the two independent clauses.

+ Gasoline prices have risen by 10 percent in the last month, and so we will drive the car less often.

3. Rewrite the sentence (if it makes sense to do so).   Turn the first independent clause into a dependent clause by adding a subordinate conjunction; then insert a comma and add the second independent clause.

+ Because gasoline prices have risen by 10 percent in the last month, we will drive the car less often.

4. Delete the comma and insert a semicolon.

+ Gasoline prices have risen by 10 percent in the last month; we will drive the car less often.

Of the four ways to correct the comma splice, sentences 3 and 4 are equally suitable, but sentence 3 reads more smoothly and so is the better choice.

The semicolon is an effective and forceful punctuation mark when two independent clauses are closely related, that is, when they announce contrasting or parallel views, as the two following examples reveal.

+ The union favored the new legislation; the company opposed it. (contrasting views)
+ Night classes help the college and the community; students can take more credit hours to advance their careers. (parallel views)

How *Not* to Correct Comma Splices

Some writers mistakenly try to correct comma splices by inserting a conjunctive adverb (*also, consequently, furthermore, however, moreover, nevertheless, then, therefore*) after the comma.

– Gasoline prices have risen by 10 percent in the last month, consequently we will drive the car less often.

Because the conjunctive adverb (*consequently*) is not as powerful as the coordinating conjunction (*and, but, for*), the error is not eliminated. If you use a conjunctive adverb—*consequently, however, nevertheless*—you still must insert a semicolon or a period before it, as the following examples show.

+ Gasoline prices have risen by 10 percent in the last month; consequently, we will drive the car less often.
+ Gasoline prices have risen by 10 percent in the last month. Consequently, we will drive the car less often.

Avoiding Run-On Sentences

A **run-on sentence** is the opposite of a sentence fragment. The fragment gives the reader too little information, the run-on too much. A run-on sentence forces readers to digest two or more grammatically complete sentences without the proper punctuation to separate them.

Run-on The Internet is a major source of information and students and other researchers correctly call it a virtual library this library is not like the collections of books and magazines that are carefully shelved always waiting for students to check and recheck them too often a website disappears or changes considerably and without a backup file or a hard copy the researcher has no document to quote from and no exact citation to prove that he or she consulted an authentic source.

Revised The Internet is a major source of information. Students and other researchers correctly call it a virtual library, although this library is not like the collections of books and magazines that are carefully shelved, waiting for students to check and recheck them out. But too often a website disappears, is under construction, or changes considerably. Without a backup file or a hard copy of the site, the researcher has no document to quote from and no exact citation to prove that he or she consulted an authentic source.

As the revision above shows, you can repair a run-on sentence by (1) dividing it into separate, correctly punctuated sentences and (2) by adding coordinating conjunctions (*and, but, yet, so, or, nor*) between clauses.

Making Subjects and Verbs Agree in Your Sentences

A subject and a verb must agree in number. A singular subject takes a singular verb, whereas a plural subject requires a plural verb.

Singular Subject	Plural Subjects
the engineer calculates	engineers calculate
a report analyzes	reports analyze
a policy changes	policies change

You can avoid subject-verb agreement errors by following several simple rules.

1. Disregard any words that come between the subject and its verb.

Faulty: The customer who ordered three parts want them shipped this afternoon.
Correct: The <u>customer</u> who ordered three parts <u>wants</u> them shipped this afternoon.

2. A compound subject (two parts connected by *and*) **takes a plural verb.**

Faulty: The engineering department and the safety committee prefers to develop new guidelines.
Correct: The engineering department and the safety committee prefer to develop new guidelines.

3. When a compound subject contains *neither . . . nor* or *either . . . or*, the verb agrees with the subject closest to it.

Faulty: Either the residents or the manager are going to file the complaint.
Correct: Either the residents or the manager is going to file the complaint.
Correct: Either the manager or the residents are going to file the complaint.

4. Use a singular verb after collective nouns (like *committee, crew, department, group, organization, staff, team*) **when the group functions as a single unit.**

Correct: The crew was available to repair the machine.
Correct: The committee asks that all recommendations be submitted by Friday.

but

Correct: The staff were unable to agree on the best model.
(The staff acted as individuals, not a unit, so a plural verb is required.)

5. Use a singular verb with indefinite pronouns (such as *anyone, anybody, each, everyone, everything, no one, somebody, something*).

Each of the programmers has completed the seminar.
Somebody usually volunteers for that duty.

Similarly, when *all, most, more*, or *part* is the subject, it requires a singular verb.

Most of the money is allocated.
Part of the equipment was salvageable.

6. Words like *scissors* and *pants* are plural when they are the true subject.

Faulty: A pair of trousers were available in his size. (*Pair* is the singular subject.)
Correct: The trousers were on sale.

7. Some Latin/Greek plurals (*curricula, data, media, phenomena, strata, syllabi*) **always take a plural verb.**

The data conclusively prove my point.
The media are usually the first to point out a politician's weak points.

8. Use a singular verb with fractions.

Three-fourths of her research proposal was finished.

Writing Sentences That Say What You Mean

Your sentences should say exactly what you mean, without doubletalk, misplaced humor, or nonsense. Sentences are composed of words and word groups that influence each other.

Writing Logical Sentences

Sentences should not contradict themselves or make outlandish claims. The following examples contain errors in logic; note how easily the suggested revisions solve the problem.

Illogical: Steel roll-away shutters make it possible for the sun to be shaded in the summer and to have it shine in the winter. (The sun is far too large to shade; the writer meant that a room or a house, much smaller than the sun, could be shaded with the shutters.)

Revision: Steel roll-away shutters make it possible for owners to shade their living rooms in the summer and to admit sunshine during the winter.

Using Contextually Appropriate Words

Sentences should use the combination of words most appropriate for the subject.

Inappropriate: The members of the Nuclear Regulatory Commission saw fear radiated on the faces of the residents. (The word *radiated* is obviously ill advised in this context; use a neutral term.)

Revision: The members of the Nuclear Regulatory Commission saw fear reflected on the faces of the residents.

Writing Sentences with Well-Placed Modifiers

A **modifier** is a word, phrase, or clause that describes, limits, or qualifies the meaning of another word or word group. A modifier can consist of one word (a *green* car), a prepositional phrase (the man *in the telephone booth*), a relative clause (the woman *who won the marathon*), or an *-ing* or *-ed* phrase (*walking three miles a day*, the student was in good shape; *seated in the first row*, we saw everything on stage).

A **dangling modifier** is one that cannot logically modify any word in the sentence.

– When answering the question, his calculator fell off the table.

One way to correct the error is to insert the right subject after the *-ing* phrase.

+ When answering the question, he knocked his calculator off the table.

You can also turn the phrase into a subordinate clause.

+ When he answered the question, his calculator fell off the table.
+ His calculator fell off the table as he answered the question.

A **misplaced modifier** illogically modifies the wrong word or words in the sentence. The result is often comical.

– Hiding in the corner, growling and snarling, our guide spotted the frightened cub. (Is our guide growling and snarling in the corner?)
– All travel requests must be submitted by employees in blue ink. (Are the employees covered in blue ink?)

The problem with both of those examples is word order. The modifiers are misplaced because they are attached to the wrong words in the sentence. Correct the error by moving the modifier to where it belongs.

+ Hiding in the corner, growling and snarling, the frightened cub was spotted by our guide.
+ All travel requests by employees must be submitted in blue ink.

Misplacing a relative clause (introduced by relative pronouns like *who, whom, that, which*) can also lead to problems with modification.

- The salesperson recorded the merchandise for the customer that the store had discounted. (The merchandise was discounted, not the customer.)
- The salesperson recorded the merchandise that the store had discounted for the customer. (The salesperson recorded for the customer; the store did not discount for the customer.)
+ The salesperson recorded for the customer the merchandise that the store had discounted.

Always place the relative clause immediately after the word it modifies.

Using Pronoun References in Sentences Correctly

Sentences will be vague if they contain a faulty use of pronouns. When you use a pronoun whose **antecedent** (the person, place, or object the pronoun refers to) is unclear, you risk confusing your reader.

Unclear: After the plants are clean, we separate the stems from the roots and place them in the sun to dry. (Is it the stems or the roots that lie in the sun?)

Revision: After the plants are clean, we separate the stems from the roots and place the stems in the sun to dry.

Unclear: The park ranger was pleased to see the workers planting new trees and installing new benches. This will attract more tourists. (The trees or the benches or both?)

Revision: The park ranger was pleased to see the workers planting new trees and installing new benches, because the new trees and benches will attract more tourists.

Words

Spelling Words Correctly

Your written work will be judged in part on how well you spell. A misspelled word may seem like a small matter, but on an employment application, e-mail, incident report, letter, or short or long report it stands out to your discredit. A spelling mistake will look careless or, even worse, uneducated to a client or a supervisor. Readers will inevitably question your other skills if your spelling is incorrect.

The Benefits and Pitfalls of Spell Checkers
Computer **spell checkers** can be handy for flagging potential problem words. But beware! Spell checkers recognize only those words that have been listed in them. A proper name or infrequently used word may be flagged as an error even though the word is spelled correctly. Moreover, a spell checker will not differentiate between such homonyms as *too* and *two* or *there* and *their*. A spell checker identifies only misspelled words, not misused words. In short, do not rely exclusively on spell checkers to solve all your spelling and word-choice problems.

Consulting a Dictionary
Always have a dictionary handy. Two online dictionaries to consult are the *Merriam-Webster Dictionary* (available online at *http://www.m-w.com/*) and the *American Heritage Dictionary* (available online at *http://www.bartleby.com*).

Using Apostrophes Correctly

Apostrophes cause some writers special problems. Basically, apostrophes are used for four reasons: (1) contractions, (2) possessives, (3) plurals, and (4) abbreviations. The guidelines below will help you to sort out those uses.

1. In a **contraction**, the apostrophe takes the place of the missing letter or letters: *I've = I have; doesn't = does not; he's = he is; it's = it is.* (*Its* is a possessive pronoun (the dog and *its* bone), not a contraction. There is no such form as *its'.*)

2. To form a **possessive**, follow these rules.

a. If a singular or plural noun does not end in an *-s*, add *'s* to show possession.

Mary's locker   the woman's jacket
children's books   the women's jackets
the staff's dedication   the company's policy

b. If a singular noun ends in *-s*, add *'s* to show possession.

the class's project   the boss's schedule

c. If a plural noun ends in *-s*, add just the *'* to indicate possession.

employees' benefits   computers' speed
lawyers' fees   stores' rates

d. If a proper name ends in *-s*, add *'s* to form the possessive.

Jones's account   Keats's poetry
the Williams's house   James's contract

e. If it is a compound noun, add an *'* or *'s* to the end of the word.

brother-in-law's business   Ms. Allison Jones-Wyatt's order

f. To indicate shared possession, add just *'s* to the last name.

Warner and Kline's Computer Shop   Juan and Anne's major

g. To indicate separate possession, add *'s* to each name.

Juan's and Tia's transcripts   Shakespeare's and Byron's poetry

3. To form the plural of numbers and capital letters used as nouns, including abbreviations without periods, just add *s*. To avoid misreading some capital letters, however, you may need to add the apostrophe.

during the 1980s   all perfect 10s
their SATs   several local YMCAs
the 3 Rs   straight A's

4. For abbreviations with periods and for lowercase letters used as nouns, form the plural by adding *'s*.

his *p*'s and *q*'s   Ph.D.'s

Using Hyphens Properly

Use a hyphen (- as opposed to a dash —) for

- **compound words**

    four-part lecture   heavy-duty machine   hand-held computer

- **most words beginning with** *self*

    self-starting   self-defense   self-regulating   self-governing

- **fractions used as adjectives**

    at the three-quarter level   two-thirds majority   three-dimensional drawing

Using Ellipses

Sometimes a sentence or passage is particularly useful, but you may not want to quote it fully. You may want to delete some words that are not really necessary for your purpose. These omissions are indicated by using an *ellipsis* (three spaced dots within the sentence to indicate where words have been omitted). Here is an example.

Full Quotation: "Diet and nutrition, which researchers have studied extensively, significantly affect oral health."

Quotation with Ellipsis: "Diet and nutrition . . . significantly affect oral health."

Using Numerals Versus Words

Write out numbers as words rather than numerals

- **to begin a sentence**

    Nineteen ninety-nine was the first year of our recruitment drive.

- **to list the first number when two numbers are used together**

    The company needed eleven 9-foot slabs.

But use numerals, not words,

- **with abbreviations, percentages, symbols, units of measurement, dates**

    17%   11:30 a.m.   70 ml
    Dec. 3, 2003 $250.00   50 K

- **for page references**

    pp. 56–59

- **for large numbers**

    3,000,000   23,750   1,714

Use both numerals and words when you want to be as precise as possible in a contract or a proposal.

We agreed to pay the vendor an extra twenty-five dollars ($25.00) per hour to finish the job by the 18th of May.

Matching the Right Word with the Right Meaning

The words in the following list frequently are mistaken for one another. Some are true homonyms; others are just similar in spelling, pronunciation, or usage. The part of speech is given after each word. Use the right word in the right context.

accept (v) to receive, to acknowledge: *We accept your proposal.*
except (prep) excluding, but: *Everyone attended the meeting except Neelou.*

advice (n) a recommendation: *I should have taken Xi's advice.*
advise (v) to counsel: *Our lawyers advised us not to sign the contract.*

affect (v) to change, to influence: *Does the detour on Route 22 affect your travel plans?*
effect (n) a result: *What was the effect of the new procedure?*
effect (v) to bring about: *We will try to effect a change in company policy.*

all ready (adj) two-word phrase *all + ready*; to be finished; to be prepared: *We are all ready for the inspector's visit.*
already (adv) previously, before a given time: *Our webmaster had already constructed the sites.*

ascent (n) move upward: *We watched the space shuttle's ascent.*
assent (n) agreement: *She won the teacher's assent.*
assent (v) to agree: *The committee asked the company to assent to the new terms.*

attain (v) to achieve, to reach: *We attained our sales goal this month.*
obtain (v) to get, to receive: *You can obtain a job application on their website.*

cereal (n) breakfast food: *Shironda always puts fruit on her cereal.*
serial (adj) arranged in sequence: *The files were in serial order.*
serials (noun) journals/magazines published at regular intervals.

cite (v) to document: *Please cite several examples to support your claim.*
site (n) place, location: *They want to build a parking lot on the site of the old theater.*
sight (n) vision: *His sight improved with bifocals.*

coarse (a) rough: *The sandpaper felt coarse.*
course (n) subject of study: *Sharonda took a course in calculus this fall.*

complement (v) to add to, enhance: *Her graphs and charts complemented my proposal.*
compliment (v) to praise: *The customer complimented us on our courteous staff.*

continually (adv) frequently and regularly: *This answering machine continually disconnects the caller in the middle of the message.*
continuously (adv) constantly: *The air conditioning is on continuously during the summer.*

council (n) government body: *The council voted to increase salaries for all city employees.*
counsel (n) advice: *She gave the trainee pertinent counsel.*

discreet (adj) showing respect, being tactful: *The manager was discreet in answering the complaint letter.*
discrete (adj) separate, distinct: *Put those figures into discrete categories for processing.*

dual (adj) double: *A clock-radio serves a dual purpose.*

duel (n) a fight, a battle: *The argument almost turned into a duel.*

eminent (adj) prominent, highly esteemed: *Dr. Felicia Rodriguez-Rollins is the most eminent neurologist in our community.*

imminent (adj) about to happen: *A hostile takeover of that company is imminent.*

fair (n) convention, exhibition: *The technological fair featured a DVD home theater with five satellite speakers.*

fair (adj) honest: *Their price was fair.*

fare (n) cost for a trip: *She was able to get a discount on a round-trip fare.*

fare (n) food: *They ate East Asian fare.*

foreword (n) preface, introduction to a book: *The foreword outlined the author's goals and objectives in her research study.*

forward (adv) toward a time or place; in advance: *We moved the time of the visit forward on the calendar so we could meet the overseas manager.*

forward (v) to send ahead: *We forwarded her e-mail to her new address.*

imply (v) to suggest: *The supervisor implied that the mechanics had taken too long for their lunch break.*

infer (v) to draw a conclusion: *We can infer from these sales figures that the new advertising campaign is working.*

it's (pronoun + verb) contraction of *it* and *is*: *Do you think it's too early to tell?*

its (adj) possessive form of *it*: *That printer is on its last cycle.*

knew (v) (past tense of *know*): *She knew the new regulations.*

new (a) never used before: *The subwoofer was new.*

lay/laid/laid (v) to put down: *Lay aside that project for now. He laid aside the project. He had already laid aside the project twice before.*

lie/lay/lain (v) to recline: *I think I'll lie down for a while. He lay there for only two minutes before the firefighter rescued him. She has lain out in the sun too often.*

lose (v) to misplace, to fail to win: *Be careful not to lose my calculator. I hope I don't lose my seat on the planning board.*

loose (adj) not tight: *The printer ribbon was too loose.*

miner (n) individual who works in a mine: *His uncle was a miner in West Virginia.*

minor (n) someone under legal age: *The law forbids the sale of tobacco to minors.*

pare (v) to cut back: *Sandoval pared the skin from the apple.*

pair (n) a couple: *They offered a pair of resolutions.*

pear (n) a fruit: *Alphonso ate a pear with lunch.*

passed (v) went by (past tense of *pass*): *He passed me in the hall without recognizing me.*

past (n) time gone by: *We've never used their services in the past.*

personal (adj) private: *The manager closes the door when she discusses personal matters with one of her staff.*

personnel (n) staff of employees: *All personnel must participate in the 401(k) retirement program.*

perspective (n) view: *From the customer's perspective, we are an honest and courteous company.*
prospective (adj) expected, likely to happen or become: *E-mail the prospective budget to district managers.*

plain (adj) simple, not fancy: *He ate plain food.*
plane (n) airplane: *The plane for Dallas leaves in an hour.*
plane (v) to make smooth: *The carpenter planed the wood.*

precede (v) to go before: *A slide show will precede the open discussion.*
proceed (v) to carry on, to go ahead: *Proceed as if we had never received that letter.*

principal (adj) main, chief: *Sales of new software constitute their principal source of revenue.*
principal (n) the head of a school: *She was a high school principal before she entered the business world.*
principal (n) money owed: *The principal on that loan totaled $32,855.*
principle (n) a policy, a belief: *Sales reps should operate on the principle that the customer is always right.*

quiet (adj) silent, not loud: *He liked to spend a quiet afternoon in the park.*
quite (adv) to a degree: *The officer was quite encouraged by the recruit's performance.*

stationary (adj) not moving: *Miguel rides a stationary bicycle for an hour every morning.*
stationery (n) writing supplies, such as paper and envelopes: *Please stop off at the stationery store and buy some more address labels.*

than (conj) as opposed to (used in comparisons): *He is a faster keyboarder than his predecessor.*
then (adv) at that time: *First she called the vendor; then she summarized their conversation in an e-mail to her boss.*

their (adj) possessive form of *they*: *All the lab technicians took their vacations during June and July.*
there (adv) in that place: *Please put the printer in there.*
they're (pronoun + verb) contraction of *they* and *are*: *They're our two best customer service representatives.*

waiver (n) international relinquishment of a right, claim, or privilege: *The company issued a waiver so that additional liability insurance would not have to be secured.*
waver (v) to shake, to move: *Our company would not waver in its commitment to safety.*

who's (pronoun + verb) contraction of *who* and *is*: *Who's up next for a promotion?*
whose (adj) possessive form of *who*: *Whose idea was that in the first place?*

you're (pronoun + verb) contraction of *you* and *are*: *You're going to like their decision.*
your (adj) possessive form of *you*: *They agree with your ideas.*

Proofreading Marks

Mark	Example	Result
⌃o	Correct a typo.	Correct a typo.
r⌃/m⌃/⌃o	Correct more than one typo.	Correct more than one typo.
t	Insert a letter.	Insert a letter.
or words	Insert a word.	Insert a word or words.
ℓ	Make a deletion.	Make a deletion.
ℰ	Delete and close up space.	Delete and close up space.
⌒	Close up extra space.	Close up extra space.
#	Insert proper spacing.	Insert proper spacing.
#/⌒	Close up and insert space.	Close up and insert space.
eq #	Regularize proper spacing.	Regularize proper spacing.
tr	Transpose letters indicated.	Transpose letters indicated.
tr	Transpose as words indicated.	Transpose words as indicated.
tr	Reorder shown as words several.	Reorder several words as shown.
⊏ ⊏	Move text to left.	Move text to left.
⊐ ⊐	Move text to right.	Move text to right.
¶	Indent for paragraph.	Indent for paragraph.
no ¶ ⊏	No paragraph indent.	No paragraph indent.
// //	Align type vertically.	Align type vertically.
run in	Run back turnover lines.	Run back turnover lines.
⌐	Break line when it runs far too long.	Break line when it runs far too long.
⊙	Insert period here.	Insert period here.
⋀	Commas, commas everywhere.	Commas, commas everywhere.
⌄	Its in need of an apostrophe.	It's in need of an apostrophe.
⌄/⌄	Add quotation marks, he begged.	"Add quotation marks," he begged.
;	Add a semicolon, don't hesitate.	Add a semicolon; don't hesitate.
:	She advised, "You need a colon."	She advised: "You need a colon."
?	How about a question mark.	How about a question mark?
(/)	Add parentheses, as they say.	Add parentheses (as they say).
lc	Sometimes you want Lowercase.	Sometimes you want lowercase.
caps	Sometimes you want upperCASE.	Sometimes you want UPPERCASE.
ital	Add italics instantly.	Add italics *instantly*.
bf	Add boldface if necessary.	Add **boldface** if necessary.
wf	Fix a wrong font letter.	Fix a wrong font letter.
sp	Spell out all ③ terms.	Spell out all three terms.
⌄	Change x to a subscript.	Change x to a subscript.
⌃	Change y to a superscript.	Change y to a superscript.
stet	Let stand as is.	Let stand as is. (To retract a change already marked.)

INDEX